Biodynamic Agriculture

Willy Schilthuis

Biodynamic Agriculture

Floris Books

Translated from Dutch by Tony Langham and Plym Peters

First published in Dutch in 1994 by Christofoor Publishers
First published in English in 1994 by Floris Books
This edition 2003

British Library CIP Data available

ISBN 086315-397-6

Printed in Poland

Toruńskie Zakłady Graficzne „Zapolex" Sp. z o.o.
ul. Gen. Sowińskiego 2/4; 87-100 Toruń
www.zapolex.pl

Contents

Acknowledgments

The author and publishers would like to acknowledge the invaluable assistance received from Patricia Thompson, Organizing Secretary of the Bio-Dynamic Agricultural Association in Great Britain. She responded patiently and promptly to what must have seemed an endless stream of requests.

Acknowledgments are also due to John Soper (Great Britain), Peter Proctor (New Zealand) and Evelyn Speiden Gregg (USA) for background material.

Photographic acknowledgments

Cover: Oaklands Park, Tyll van der Voort; The Biodynamic Agricultural Association (UK) pp. 19, 20, 29, 30, 32, 34, 36, 49, 50, 61, 65, 66, 68, 69, 70, 71, 72, 73, 75, 76, 79, 82, 84, 85, 86, 87, 92, 100, 104, 107, 109, 110, 111, 112; Botton Hall, Camphill p. 98; Getty Images pp. 11, 16; Goetheanum Library p. 43, Verlag am Goetheanum p. 54; Maria Thun pp. 8, 18, 89; Soil Association p. 40.

The traditional face of agriculture

~ 1 ~

Agriculture and the environment

Over the last hundred years, and especially in the second half of the twentieth century, agriculture throughout the western world has endured a time of enormous changes and difficulties. Slowly but surely, powerful machines have all but replaced collective human labour along with such traditional agricultural aids as the horse. At the same time, there has been a rapid deepening of knowledge about the chemical composition of plants, the chemical nutrients which plants need, and the ways in which these can be added to the soil. All this new chemical knowledge (pesticides, herbicides, fertilizers, plant foods), and engineering expertise (machines and computers) has been in the general service of agriculture since the Second World War.

As a result, modern farmers have been offered a vastly greater range of possibilities. Where natural conditions imposed limitations, these could more often than not be overcome with the help of the new technologies.

Only in recent years have the problematic consequences begun to emerge. In Europe alone, high yields brought about huge excesses of grain production, and similarly too much milk, meat and olive oil. Phrases such as 'butter mountain,' 'wine lake,' have passed into common language, together with their bureaucratic solution, 'set-aside,' the latest addition to agricultural Euro-jargon. Underlying these problems is the excessive use of chemical preparations of all sorts: fertilizers, pesticides and herbicides. Far from our imaginings of the rural idyll, modern agriculture — along with industry and road traffic — has proved to be one of the great polluters of the earth, water and atmosphere.

Industrialization

After the Second World War, government policies in virtually all the 'developed' countries were focused on industrialization. Industry expanded because there was a great demand for industrial products in the re-building period after the destruction of wartime.

Backed by the huge investment of the Marshall Plan, it was possible for reconstruction to take place in a relatively short time. There was work for everyone and standards of living rose rapidly. The farming sector of the population was keen to share in these higher standards of living and higher salaries. Farm labourers went to work in factories or in the building industry, or they asked for more pay. For many farmers, these increased pay demands were out of their reach. All these factors contributed to the gradual replacement of manual labour by machines.

As the process of industrialization also affected agriculture, people even started thinking about agriculture as though it, too, were a mechanized industry. A properly managed industrial enterprise aims to use the machines required in the production process as efficiently as possible, because they are expensive to purchase and must therefore be utilized to the greatest productive use. In agriculture the use of expensive machines had the same effect.

Many farmers decided to specialize: either in arable farming with the cultivating implements involved in that, or in dairy farming with modern milking parlours. Those who work the land with machines like to have large, straight fields, in order to work fast and efficiently. Inevitably, then, large-scale arable farming became the general aim, and it soon became clear that the smaller mixed farms were unable to survive the

An automatic milking machine

economic pressures. In Britain, another major factor in specialization during the post-war period has been the system of grants awarded by the Ministry of Agriculture.

Intensive livestock farming

In arable farming, as we have seen, modernizing and industrializing came about through mechanization and an increase of scale. But this was not universally possible. Where the farms tended to be on a smaller scale, and especially in areas with sandy soil which is not ideal for arable farming, another solution was found: intensive farming. Small farms were able to begin specializing in a particular branch of animal husbandry on the available acreage, for example, pig breeding, calf rearing or chicken farming.

Obviously a small farm cannot grow the food for all these animals, but this is not a problem as the world market prices for raw materials such as corn, soya, cassava, shredded coconut, groundnuts and so on, are very low and these can be bought ready packaged from feed merchants. In this way many small farmers were able to keep going, although the investments required for intensive farming in this way were almost

The nitrogen cycle

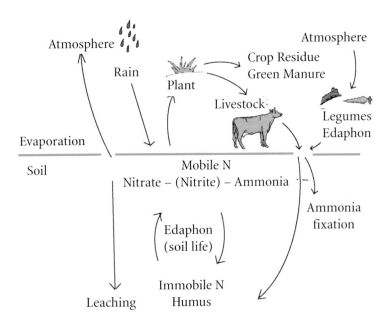

impossible to service in many cases, so that the farmer became completely dependent on the bank and/or the feed merchant.

This whole process of the industrialization of agriculture was strongly supported by the various different EC governments and the price policies which they implemented. In the information which was made available, the research that was carried out and the credit facilities provided, specialization and mechanization were encouraged everywhere. Any negative side effects were often dismissed as being trifling or were ignored entirely — until the enormous problem of agricultural surpluses could no longer be afforded and the pollution caused by fertilizers could no longer be denied.

The role of nitrogen

These problems of agricultural surpluses and pollution by fertilizer centre on the role played by nitrogen (the element's chemical symbol is N), in its various forms.

The air around us consists of as much as 78% nitrogen, but it is inert, and not active. It does not have any chemical effect and is not beneficial for plants. There is virtually no nitrogen in the Earth's crust and the rock of which it is composed. Nevertheless, all living creatures need nitrogen in order to build up the protein in their body.

In nature, nitrogen is absorbed from the air by certain plants, together with bacteria, and by some bacteria in the soil and algae so that it becomes part of the cycle of living organisms. The plants which are well known for this capacity to absorb nitrogen, and which are therefore widely used, belong to the Papilonaceae family, including every variety of clover. Species of this family and some bacteria in the soil absorb the nitrogen present in the air and the soil and assimilate it. When these plants are then eaten by animals or when they rot down, the nitrogen stored in them becomes available to other plants and to the lower organisms in the soil. Animal manure contains approximately 0.4% nitrogen. Obviously plants also require other nutrients from the soil, but these are found naturally in the soil and not in the air. The soil is actually a finely tuned eco-system of micro-organisms, plant roots and lower organisms, in which there is a constant exchange of substances, and all sorts of biochemical processes are taking place.

This process means that, left to nature, the amount of nitrogen available in the soil is limited because it must first be absorbed from the air by plants. The major event in our own century is that we now have the technological means to extract nitrogen from the air and therefore have an almost unlimited abundance of artificial fertilizers based on nitrogen. The unforeseen consequence is that accumulating excesses of these different chemical forms of nitrogen have disturbed the ecological balance: in the soil and in water as nitrates (NO_3); as ammonia (NH_4) in the air as well as the soil. These excess nitrogen compounds in the air are largely responsible for so-called 'acid rain,' with its disastrous effects on forests and the aquatic life of streams, lakes and rivers. Because of human technological skills, the natural cycle involving nitrogen has come to be completely disrupted, with all that that entails.

Herbicides and pesticides

The excessive use of nitrogen, phosphorus and potassium (NPK) in the form of artificial fertilizers has greatly increased agricultural output. When agriculture became an industry, one of the main objectives was 'higher yields.' This was successfully achieved. In 1965, one hectare of land yielded approximately 4500 kg of wheat. In 1990, the same area of a clay soil could yield 8000–9000 kg.

These results are not only due to chemical fertilizers but are also the result of the development of new, higher yielding varieties of wheat. Comparable results have been achieved for virtually all crops, albeit at the expense of the resistance and the health of those crops.

The tremendous rise in the use of herbicides and pesticides and the expansion of the related chemical industry are inextricably linked to the increased production resulting from the use of nitrates. Nitrogen makes plants grow and increase in weight but this change is not always balanced by a corresponding increase in the plants' strength, structure and resistance. As a result, aphids, caterpillars and fungi see such plants as good hosts. Huge amounts of poison are therefore sprayed on plants and fields to combat these 'diseases,' and not only to combat diseases, but also to eradicate weeds.

In the past, weeding in the traditional way by hand, using a hoe, was an extremely laborious, time-consuming and backbreaking job for farm and market-garden workers. It was an enormous saving when chemicals were introduced to do the work. But this had an extremely negative effect on the living organisms in the soil, and ultimately on ground water and the environment as a whole. The original optimistic reassurances that the chemicals would in time become less toxic and would break down in the soil did not turn out to be true. Many chemicals were prohibited after being used for several years because they were not completely degradable and furthermore appeared to have carcinogenic (cancer-producing) side-effects.

The landscape

The industrialization of agriculture led to a steady transformation of the landscape. Miles of hedgerows, coppices and field ditches were removed or filled in. Land drains were installed with the help of government subsidies. A lower water table meant that farmers could start using their increasingly heavy and large tractors on the land earlier every spring.

A result of agricultural water pollution

These practices soon gave rise to considerable anxiety among those concerned with nature conservation. They accused farmers of not caring about natural life and the landscape. In turn, the farmers accused these nature conservation groups of not understanding the realities of agriculture as an economically independent kind of enterprise. They argued that one cannot expect a farmer working alone or with few farmhands, to maintain hedgerows and coppices which have become completely redundant, simply because town-dwellers find them charming and beautiful. Though a certain consensus has been reached in recent years with the onset of 'green' politics, the farming community and the nature conservation groups have tended to view each other with mutual suspicion and hostility.

Water pollution

But perhaps the most important accusation levelled against modern agriculture has found little response. Farmers were held responsible not only for the destruction of the landscape, but also for serious pollution of the environment. These claims have been well supported by thorough research, including water studies in relation to ditches and streams, and the surface water.

Pollution of surface water is caused by our whole society, which produces unimaginable amounts of waste matter. In the past small amounts of organic waste matter in the water were cleaned up by the living organisms in the water (algae, bacteria, fish), but this is no longer possible. In the first place, there is much too much waste to be dealt with. In the second place, more and more of the waste matter is not organic, but non-biodegradable waste matter which also contains heavy metals.

The pollution of water caused by agriculture is the result of two main components, namely, herbicides and pesticides, and fertilizers. Where a large number of fish suddenly die in a ditch or a stream after a farmer has been particularly heavy-handed with a certain chemical or has simply cleaned a sprayer in the

Spraying fields

ditch, it is usually possible to trace the guilty party. It is much more difficult to establish the source of water polluted by the use of too many fertilizers. If a field or a meadow is liberally fertilized and there is then a period of rain, or even just one downpour, some of the fertilizer is washed away through the drainage system and ends up in the surface water. In this case the water has been fertilized. The commonly used term for this is the eutrophication of the surface water. The result is that the balance of the eco-system of the surface drain or ditch is seriously disturbed.

It has now become clear that pollution resulting from fertilizers and herbicides and pesticides is not only excessive in the surface water, but has also penetrated the ground water. In a number of areas where drinking water has been pumped up, it has proved to be unsuitable for consumption because of the presence of herbicides, pesticides and nitrates from fertilizers. Fertilizers, which were once the farmers' greatest boon and a pre-condition for a good harvest, have now become one of the greatest environmental pollutants.

It has become increasingly clear that modern industrial and intensive farming techniques cause serious problems of pollution and potentially catastrophic imbalances in the natural biosphere. However, it is by no means clear how these issues should be solved. Is it possible for modern agriculture throughout the world to turn back from the path it has taken? Government measures and legal regulations have attempted to reduce the most seriously damaging consequences but as yet the results are slight.

If we are to avoid worse degradation of our environment, it is time to start thinking quite differently about agriculture in the living world. Prevailing techno-scientific methods should be progressively abandoned and replaced with a more sensitive science. The analytical, quantitative approach to plants, the soil and even animals has been very fruitful in providing a

Spraying by plane.

A field of valerian.

narrow understanding of their chemical composition, but it has not provided wider knowledge in the field of the interrelationship of living organisms, cyclical processes and the ecological balance.

Biodynamic farming provides the basis and practice for a different approach.

$\sim$ 2 $\backsim$

The Earth, a living organism

Another way of approaching the Earth, its plants and animals can be achieved simply by looking and thinking from a different perspective. Suppose we stop thinking merely in terms of litres, tons and metres of the products involved in agriculture, but think instead in terms of the life processes and ecological interrelationships. This approach leads us to recognize the organic character of the most comprehensive eco-system that we know, that is, the Earth itself.

Our globe and its surrounding atmosphere make up one large and coherent organic whole. The interrelatedness of everything within this whole has become especially clear to us in recent years as we have become aware of the widespread effects of pollution: for instance, animals in polar regions, far from our population centres, have non-biodegradable poisons (DDT) and other toxic substances in their bodies.

Enormous masses of water are constantly moving around the Earth in the currents of the oceans. There are currents from the equator to the poles, and vice versa. The water is heated up at the equator and is then moved as warm ocean currents by the eastern trade winds, in the Atlantic Ocean initially westwards at a speed of 1.2 km per hour. On the east coast of Brazil the current curves to the north, flows partly through the Gulf of Mexico and is then compressed through the Straits of Florida, moving to the east at a speed of 9 km per hour. It flows back into the Atlantic Ocean and branches off in different directions. The main stream, the Gulf Stream, moves at a speed of 3.6–5 km per hour to Europe, where it clearly influences the climate. In the Pacific Ocean, as in the Atlantic, there is a clockwise current in the northern ocean and an anticlockwise current in the

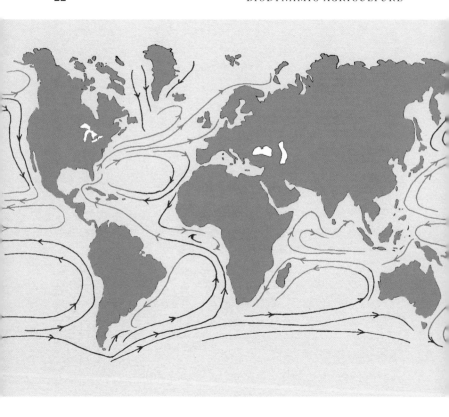

Ocean currents of the world

southern ocean, with a west-east counter-equatorial current separating them. The Indian Ocean also shows a similar tendency. In this way, the waters of the oceans of the Earth are in constant movement, linking all the continents together.

A similar process takes place in the air. The trade winds in the tropics move enormous quantities of air across the surface of the earth. From equatorial regions the warmed air rises and moves to higher latitudes in the upper atmosphere to descend again as cool dry air in the desert regions. In temperate zones there are constantly changing high and low pressure areas with their winds. Gas which is produced and rises somewhere on the Earth's surface is soon dispersed across huge areas. There is also a cycle from water to water vapour — clouds, rain, snow and hail

Satellite image of the world

The Sun with sunspots

— which then falls to earth to accumulate in streams and rivers flowing back to the sea.

Then there is the carbon dioxide cycle (CO_2) which is absorbed by plants so that they can grow and is then released when the plants decay. In the meantime, the carbon dioxide has served a useful purpose in the production of food for humans and animals, for all plant matter consists of some combination of oxygen and hydrogen with carbon (C).

These days much too much carbon is emitted in the form of CO_2 from burning fossil fuels (which were once plants absorbing CO_2). As a result the atmosphere has to absorb far too much carbon dioxide and the natural balance of gases in the air is altered giving rise to the greenhouse effect.

The Earth is like a living organism with growing plants and

decaying plants, currents of water and currents of air, which also interreact with each other. When we ask ourselves how these processes are maintained, it quickly becomes clear that the sun does this. The sun is a source of heat, light and energy, and the Earth moves around it. The processes of rain and evaporation, day and night, heat and cold, summer and winter — the sun is the source of all of these. The moon, as a satellite of the Earth, is particularly influential on the movement of water. The water on the Earth's surface reacts to the moon's attraction. With the rotation of the Earth on its own axis, the waters on the surface of the Earth are pulled towards the moon or away from the moon in a constant ebb and flow.

The dynamic force of a vortex in a shell

The plant world

All these cyclical processes which take place on the Earth are inconceivable without the Earth's cover of plants. It is the plants which are responsible for the carbon cycle by means of their process of assimilation. The seaweed in the oceans is also important in this process. Plants, interacting with bacteria, are also responsible for the nitrogen cycle, which as we have already seen in Chapter 1, has now been disrupted by the excess of technologically produced nitrogen.

The cycle of water and water vapour also takes place to a great extent through plants because water is absorbed through their roots and then evaporates through the leaves.

Looking at the eco-system of the Earth in this way, it becomes clear that a bare earth with no plants could not function as a living organism. Equally, plants without earth cannot survive. The earth system, through its protective, living, breathing cover of plants, can absorb sunlight and convert it into plant matter. The roots of the plants are able to penetrate the Earth's crust, absorb minerals from it and then enrich the soil with their own plant matter. In this way plants are able to create a middle ground between the dead mineral Earth's crust on the one hand, and the light, sun-warmed air, on the other.

Life takes place in this middle ground: the life of plants, animals and human beings. Plants convert the water and the minerals from the soil, combining them with carbon dioxide and light from the air to form living matter. In this way they create the basis for all life on Earth — not only because they are a source of food, but also because they provide protection against heat and cold, because they make the climate milder, and because they absorb rainwater and retain it in the soil. They provide us with wood for building materials and fuel, and constitute the constantly renewed layer of soil which sustains all life.

What is life?

If we want to continue to live on Earth together, we will have to look after life on Earth and maintain it; healthy, vital life processes which maintain themselves in an ecological balance, in a constant reciprocal interaction and cyclical process. What is this life?

We still do not really know. We know a lot about the nutrients which living organisms need. This knowledge is important for farmers, as without an appreciation of these substances, their chemical properties and chemical applications, modern farming would not be possible. However, biodynamic farming methods take into account not only these substances, but also that other component: life.

In biodynamic farming, we use the term life forces or etheric forces. The difference between a lump of stone and a plant is that the stone is a purely material object and only has a physical body, while a plant is of a higher order, that is, the order of life forces or matter controlled by life, which makes it something

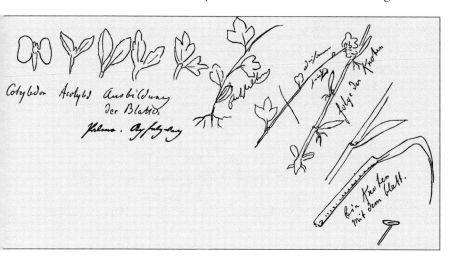

Goethe's sketches of leaf-like parts of the plant.

entirely different: a living organism. When the plant loses life, its matter is once again subject to material laws.

It is a characteristic of life that it can only be passed on by living beings. A living being can be created only if it is given life by its parents, whether these are plants, animals or humans, or even bacteria or other single-cell organisms. This life is passed from the parent to its offspring as an 'etheric body' or 'life body'.[1]

In biodynamic farming, in order to adopt the correct approach to living beings, crops, animals and the micro-life in the soil, a deeper insight is needed into the four ways in which these etheric forces or life forces operate. These are:

— the effect of heat;
— the effect of light and air;
— the effect of liquids;
— the effect of solid matter (that is, minerals.)

The life forces which have their origin in the effects of the sun, light/air and heat, affect the plant particularly through the leaves and flowers. The life forces relating to the building up of the plant from the soil, water and plant nutrients, affect the plant through the root system in the earth. At the same time, no living organism, not even bacteria, can exist without the co-operation of all four life forces. These four forces operate together on the life body of living beings. This is visible and tangible in the effect of the four elements: earth, water, fire and air, as they are often described in ancient books. Nowadays we use the terms light, heat, water and plant nutrients in practical agriculture and horticulture.[2]

As farmers and market-gardeners, our aim is to produce large numbers of healthy plants. We do this by ensuring that the plants have sufficient water and nutrients. However, if we neglect the plants' life forces, the growing crop will not have enough strength to survive and will become an easy target for parasites.

In present-day, technically oriented agriculture, it is very difficult to farm without the use of chemical pesticides and herbicides. Because not enough attention is paid to the life aspect, or the etheric forces, plants do not have sufficient vitality and are unable to build up resistance to disease. Consequently they become vulnerable to outside influences. Added to that, their value as a source of food is not as good as it could be, for the vitality — or life force — of the food that we eat, is just as important as the material substance we ingest.

The soil

Obviously the task of the biodynamic farmer is to ensure that his crops are given the chance to build up a vital life body, and consequently a strong and healthy plant body. This is achieved in particular by feeding and looking after the soil. When Rudolf Steiner, the Austrian philosopher and scientist, gave a series of lectures to farmers in 1924 which were the foundation of biodynamic agriculture, he repeatedly stressed that life can only be passed on by the living. In other words, cows and plants and the soil must be fed with living substances, or at least with nutrients

Checking the soil

Biodynamic soil (left) compared to standard soil (right)

which were created in a life process. The mistake in modern agriculture is that the soil is seen as a great store of chemical compounds from which the plant can take what it needs. Therefore, fertilization consists of applying chemical compounds: phosphates, potassium salts and nitrates. This idea, taken to its logical conclusion, has resulted in cultivation on rockwool. Rockwool is a completely dead synthetic material but saturated with water including precise dosages of dissolved nutrient minerals, it allows the horticulturalist to grow vegetables in totally artificial conditions in greenhouses. Extreme hygienic 'barrier' measures are needed to prevent infections, such as disinfecting your shoes on entering the area. Thus such systems have very little genuine relationship with farming the earth.

On the subject of fertilization, Rudolf Steiner said that we should fertilize the soil, and not the plants. This starting point is clearly diametrically opposite to that leading to the rockwool cultivation method.

If the soil, the ground in which the plants grow, is fed with organic manure and compost, it will start to live. Countless organisms can live in it, from earthworms to bacteria and fungi. This means that it is easy for the plants' roots to work their way into the soil and oxygen can penetrate everywhere. It is interesting to note that lower organisms such as parasites do not gain the upper hand in this soil.

Thus for the biodynamic farmer, looking after the soil is of primary importance, and this also involves the effect of the four life forces on the soil. There must be enough water, but not too much, for then there will be a lack of air. Can enough heat penetrate into the soil and what is the position with regard to the supply of plant nutrients? These will not be given in the form of artificial fertilizers, but in all sorts of organic manure and sometimes in ground minerals. Caring for the soil also involves tilling it at the right moment, ploughing a deep or shallow furrow, cultivating, harrowing, rolling, the correct order of crops in a crop rotation plan, mulching, and so on.

Animals

So far we have referred to the earth, water, air, heat and plants. Animals are also part of all this. However, they have a very different mode of life and therefore a different place in the interrelated system. They are not rooted in the earth, but are able to move about freely. They do not live on minerals and water and light like plants, but need living matter as food in the form of plants or other animals. There is another big difference: they have instincts, passions, feelings of hunger and thirst. They are able to express these feelings and take steps to satisfy their needs. In other words, animals not only have a physical body and a life body like plants, but they have a higher organization which enables them to act in a more independent and free fashion. They can express their feelings through their behaviour, they can make us hear their call; they have an inner world and can clearly react to and even affect the outer world, digging a hole or building a nest. Therefore, they are less dependent on the earth, less rooted in the life body of the earth. This organization, which gives animals their consciousness and the potential to express themselves, is known as their 'astral body' or 'sentient body'. It is called 'astral' because the forces which act on the astral body come from the world of stars and planets.

Every animal has its own life body and its own inner world of feelings, or astral body. However, all animals of a single species are interrelated in an even higher organization, as revealed in their behaviour and instincts. All hares have the same instincts and behave in the same way, as do all deer and all foxes. Sometimes this instinctive behaviour is less strong in domesticated species, when the human influence is dominant.[3] Nevertheless, our domestic animals and the livestock on a farm all clearly have their own species-bound behaviour. Cows walk differently, behave differently and graze differently from sheep or horses. Goats are different again.

Although these animals may graze in the same field under the same conditions, their manure will have a completely different

Sturts Farm, Hampshire

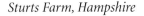

composition. The grass which they eat is assimilated and processed in a very different way in a cow and in a sheep. Cow manure is soft and slimy and flows down to form a cow pat. Sheep droppings appear in the form of solid, round pellets and these droppings smell very different from cow pats. The effect of these different types of manure on the soil is also very different. When an animal digests fodder, the substances and life forces assimilated from the plant nutrients are useful and necessary for it. What is excreted has been worked upon by the animal, with its astral body. Astral forces are contained in the manure, and these are expressed, for example, in the form of the odour and properties of the manure.

The soil therefore receives very new and valuable additions from this manure. The soil is enriched by these astral forces and is therefore better able to develop its own soil life, to form humus and to become more open to the beneficial effects of the sun, the moon and the other planets on the plants.

The farming cycle

Animals ensure that there is a cycle of substances on a farm. What livestock, for example, eat as fodder and then digest, later becomes partly available in the form of manure. The manure is returned to the land and the land can then produce more fodder and other products. This cycle of substances is permeated with life forces and qualities and this gives the farm its own individual quality.

Therefore the ideal set-up for a biodynamic farm is a mixed farm with livestock and fodder crops, producing its own manure for the soil. This farm is (or should be) a self-support-ing closed organism, in which the different organs such as the fields, the manure, the meadows and the animals all interrelate properly. One organ should not produce more than what is needed by the other organs, except of course for what is pro-duced for human consumption. Foreign elements, such as arti-ficial fertilizers or bought-in feed, which do not belong in this organism, are not constantly introduced on this farm. Elements

Spreading manure

are assimilated from the air such as the nitrogen which is absorbed by leguminous plants, the carbon dioxide which is assimilated and the energy of the sun. The farm feeds on these and the food produced for man is created as a contribution to life on Earth.

Biodynamic preparations

However, there is an additional aspect which serves to support the effect of the astral and life forces. Because the vitality of plants has deteriorated in our century, the conditions under which plants have to grow are becoming increasingly difficult, and the soil which provides nutrients for the plants is gradually becoming poorer and unhealthier. Special compounds are used in biodynamic farming to strengthen and restore the health of the soil and the plants. It is a matter of giving the life forces and astral forces better access to the life processes in the soil, the composting process and the plants. This is achieved by making use of very ordinary raw materials such as cow manure and quartz crystal, and of plants such as dandelions and yarrow. These are treated in a special way to prepare compounds. This will be described in further detail in Chapter 4, which deals with arable farming and compost-making. The effect of these preparations is comparable to the effect of a homeopathic remedy, in so far as very small amounts are involved. These are sprayed on the soil or the plants, or added to the manure and compost-heaps.

The farmer and the farm organism

Nowadays the term 'agricultural manager' is sometimes used in farming. The word 'farmer' can in some contexts have connotations of 'stupidity' and 'conservatism.' For a long time biodynamic farmers were accused of being conservative, wishing to retain old values and resisting modern innovations. This was not entirely without justification. It certainly applied to the use of chemical weed killers and chemicals used for combating disease. Although there are in general few problems with plant diseases and the chemicals used to combat them were not really missed, there were — and still are — a few problem diseases.

A great deal of attention must be devoted to weed control, and this requires some expertise. It is a matter of rotating crops correctly, working the soil in the right way and at the right time. In general, it can be stated that the biodynamic farmer must be an expert who has to make the right choices and the right decisions on his own land, depending on the conditions of his own farm. He cannot rely on general rules.

The farmer, with his work force, manages the farm as a unified organism comprising the whole environment of the farm, including the trees and the hedgerows along the roads and ditches. Birds and insects also play a role within this organism. In order to create the conditions required for the survival of

A farmer making compost

insects such as bees and butterflies, there must be spots where wild plants can grow and flower, for example, along riverbanks and in copses.

It is an art to impose the correct balance between the different living parts of this organism: the arable fields, the meadows, the livestock, the ditches and the copses, the use of preparations and composting manure.

Nowadays biodynamic farmers are no longer considered backward or conservative. They are seen as pioneers or innovators of farming methods for the future, allowing more room for the environment and nature. They are in the forefront of innovations in agriculture and the abolition of toxic and chemical compounds. Very slowly their views are being heeded, and their methods and results are attracting interest.

～ 3 ～

Biodynamics: a brief history

We think of agricultural concerns as being a creation of our modern age, and look back only a few years to what seems like a golden age of tranquil and well-balanced farming. However, farmers and large landowners were already concerned about the situation in agriculture in the second half of the nineteenth century.

Agricultural problems in the nineteenth century

From the 1850s onwards, agriculture throughout the western world was confronted with major problems. The population of central Europe was growing and increasing amounts of food were required. However, the fertility of land which had been farmed for a long time was declining. Farmers in Holland, to take only one example, were particularly worried about areas of sandy soil where they could no longer grow rye satisfactorily, for reasons which could not be explained at the time. In Germany, Poland, England, Ireland, Sweden and Holland, great numbers of rural families emigrated to America because they simply could no longer survive off their own lands.

The answer to some of these problems was finally provided by science, particularly by chemists. The name of Justus von Liebig (1803–73), 'the father of agricultural chemistry,' is especially well known in this context. He studied plants to find out what substances they need from the soil in order to grow and in what form these substances are absorbed. Knowledge which is considered self-evident nowadays and is taught at every agricultural college, was the subject of pioneering research at that time. Phosphorus and potassium were shown to be important substances and

Justus von Liebig (1803-73)

the addition of nitrogen in particular helped plants to grow. Animal manure contains nitrogen, but supplies were limited. This led to the importation of guano from Peru and saltpetre from Chile as sources of nitrogen, though only limited quantities were used because of the distances and costs involved. Obviously farmers, being used to stable manure from their own livestock, were at first very suspicious of these strange chemicals. Then the First World War broke out in Europe and the post-war period was to bring enormous changes to the agricultural scene.

At that time the first steps were taken towards a chemical and technical form of agriculture. Wartime technology had discovered a large-scale method of making explosives with nitrogen extracted from the atmosphere, and after the war this nitrogen became easily and cheaply available in the form of saltpetre (NO_3). The war had also seen the rapid development of tanks, and these provided the technology for prototype tractors. The manufacture of the poison gases used in trench warfare gave rise to the chemical know-how for the production of insecticides.

Movements for biological agriculture

Against the background of all these developments, there were farmers who were concerned to note, for example, that the germination of seeds had deteriorated and that problems arose getting cows in calf. These and similar difficulties had no immediate 'chemical' solution, but called for deeper examination. Simultaneously in Germany and in a number of other countries, biological farming methods were being examined and developed quite independently. On the other side of the world, in India, Sir Albert Howard (1873–1947) carried out research into plant diseases in the crops being grown there by British plantation owners. He discovered a relationship between the extent of disease in a crop and the composting methods and treatment of the soil. On the basis of his discoveries, he developed a composting system which was used by planters in many colonial countries.

Lady Eve Balfour (1898–1990), living in England at the same time, was convinced that the only healthy method of farming was a cyclical method of rotation. She applied this system consistently on her own farm in Suffolk without the use of artificial fertilizers or cattle feed, and adopted Howard's composting system. Her account of her work in *The Living Soil* (1943) gave rise to such widespread response that, after the

Sir Albert Howard   *Lady Eve Balfour*

Second World War, she founded the Soil Association and the magazine *Mother Earth.* Subsidiary organizations were founded in other English-speaking countries and the Soil Association still plays an important role in the international movement for organic agriculture.

The ideas of Howard and Lady Balfour inspired many people worldwide to develop their own ideological and practical farming systems, under the various modern designations of biological, organic or ecological farming. From the 1950s up until the 1970s, national movements with more or less their own characteristic features arose in different countries. For example, the Rusch-Müller method or bio-organic method, which uses the guaranteed brand name Bioland, was established in Switzerland and Germany. In France, there is a fairly large organization, Nature et Progrès. In the last twenty years there has been a strong co-operative venture between these organizations in different countries. The International Federation of Organic Agriculture Movements (IFOAM) was founded in 1972 with worldwide contacts. Even in countries with little interest in biological farming, groups can become members of IFOAM and acquire a great deal of knowledge as well as technical advice in this way. The biodynamic movement is actively involved in this international umbrella organization, though it clearly retains its own character and ideas.

Rudolf Steiner and the Agricultural Course

Rudolf Steiner (1861–1925) grew up among peasants in the countryside of Austria. He had a strong connection to nature as a child and later went to study sciences in Vienna. He made a special study of Goethe's scientific writings and went on to help edit them in a new collected edition brought out by the Goethe archives in Weimar.

Steiner was aware that natural science did not provide the whole picture of what was going on in the physical and living world. Even as a boy, he knew from his own experiences that beyond the physical world there was another world which could be 'seen' and 'heard' but not with physical eyes and ears. He experienced contact with the spiritual beings that lay hidden 'behind' the physical world. But there were few people with whom he could share these experiences.

Rudolf Steiner

During his twenties, Steiner became convinced that the methods of modern science can understand only what is dead in nature, because it overlooks the spiritual aspect of reality. He saw that Goethe's natural scientific writings built a bridge between nature and the spirit. He went on to introduce a 'spiritual science,' which he called anthroposophy, which gave rise to many new ideas and impulses in a number of fields. These included very practical initiatives in education, health care, the care of the mentally handicapped, and so on. People from many different specializations turned to him for inspiration and new ideas, including a number concerned with agriculture. Some of the farmers, large landowners and agriculturalists turned to Rudolf Steiner to ask him whether he could offer them some new insights for healthy farming aimed at the future. In June, 1924, Steiner delivered a series of lectures to about a hundred farmers on the Koberwitz estate near Breslau, which was then in the eastern part of Germany, and is now Wroclaw in Poland.

The course in Koberwitz was organized for farmers and agriculturalists who had joined the Anthroposophical Society. In this course of eight lectures Steiner presented his spiritual and scientific views on the relationship of nature and agriculture and on the development of agriculture. These lectures, known simply as the Agricultural Course, still form the basis of biodynamic farming today, and are published under the title: *Agriculture* (see Further reading).[4]

However, spiritual and scientific ideas do not in themselves constitute a practical method of farming, and as both Steiner and the farmers at the course realized this, it was decided to establish the so-called *Versuchsring* (Experimental Circle), a group of farmers who would implement these ideas and evaluate them. This experimental group was supported by the Faculty of Science at the Goetheanum, a research institute founded by Steiner at Dornach in Switzerland.

Steiner died in 1925 but his ideas were already being enthusi-

Ehrenfried Pfeiffer

astically put into practice in several European countries. A report dating from 1928 shows that at that time there were sixty-six farms based on biodynamic principles and the Experimental Circle had 148 members. There were working groups of farmers who were concerned with practical questions, such as the use of preparations, and who also provided information for other interested farmers.

One of the researchers in Dornach was Ehrenfried Pfeiffer, a biochemist and close associate of Steiner's. In 1926, the first biodynamic farms were established in Loverendale, Holland, and Pfeiffer was appointed as director. He divided his time between his research in Dornach and his work at Loverendale. In 1940, he went to the United States to introduce biodynamics there. He founded a research laboratory in New York state which is still supported by the Pfeiffer Foundation. He also set up a pilot scheme in California to convert city waste into biodynamic compost. In his capacity as advisor for the US Department of Agriculture, he successfully helped combat foot and mouth disease. Pfeiffer is also the author of several widely available and much read books (see Further Reading).[5]

The search for quality

It was not only the farmers who were concerned with biodynamic farming, but also consumers. While one important consideration is the fertility of the earth, and for farmers this is the main reason for using biodynamic methods, directly related to this is the quality of the food produced. Farmers in the early 1900s, while noticing problems with regard to the health of their cattle, also noted that there was a deterioration in the quality of the food produced.

During the same period, consumers were beginning to take an active interest in the way in which food was produced. There emerged various movements in the field of health and nutrition, such as vegetarianism and the nutritional ideas of Bircher-Benner.

The anthroposophical movement showed a great deal of interest in biodynamic products. In Germany, the brand name Demeter was introduced in 1928. This was to give consumers the certainty that they were really buying products from a biodynamic farm which was affiliated to the biodynamic Demeter organization. This Demeter symbol was the pioneering label of all the brands and logos for organic and 'green' products which are increasingly familiar to us today. In due course, a magazine was published to disseminate information, and farm visits were organized to establish contact between producers and consumers.

Over the next ten years or so, much work was achieved in successfully establishing biodynamic farming. However, as the National Socialists progressively took over political power in Germany, further development became more and more difficult. In the end biodynamic farming and the entire anthroposophical movement were banned, a ban which also applied in the countries occupied by Germany during the Second World War.

A new start

When the war ended in 1945, there was virtually nothing left of the biodynamic movement. Germany, where biodynamic farming had originated, was in a very bad way. East Germany, where the greatest number and the largest farms had been based on biodynamic methods, became part of Poland. Small centres had survived here and there in West Germany, the Netherlands, the Scandinavian countries and England.

However, in 1946 the *Forschungsring für biologisch-dynamische Wirtschaftsweise* was re-established, with Hans Heinze as the organizer and co-ordinator. The magazine *Lebendige Erde* (Living Earth) was soon being published again in small numbers. In 1954, the Demeter organization was re-established, so as to renew the aim of producing a guaranteed brand name for biodynamic products.

New initiatives also took place in other countries. Once again, associations were founded, and international contacts were restored. In Holland, a biodynamic association, founded in 1937 but which had operated underground during the last years of the war, started up again. The association consisted mainly of interested people and consumers but did not include many farmers. Later, in the post-war years the wider response of consumers and farmers in general was very limited. There was such an insistence on the need to produce large amounts of food quickly and on the new technological possibilities, that hardly any attention was paid to quality or to the consequences for the natural environment.

An international movement

The beginnings of biodynamics in Great Britain can be traced back to 1928. In that year, D.N. Dunlop, who had met Rudolf Steiner during his visits to England, organized a meeting to which he invited experts to give accounts of activities which had arisen out of anthroposophy on the European continent. Carl Mirbt (who later changed his name to Mier) came as a representative of Count Keyserlingk to speak about biodynamics. Dunlop was so impressed that he invited Carl to return the following year with his family, and make a start.

Carl Mier began working with Marna Pease on her estate in Northumberland. Soon they all moved to Bray, in Berkshire, where the garden became transformed through intensive use of the preparations, and was visited by gardeners from all over the country. Mrs Pease became the leader of the Anthroposophical Agricultural Foundation, a post which she held until 1946 when David Clement took over from her. He and his wife were farming at Broome Farm in Clent, where they had been since 1940. This was run to supply the nearby Sunfield Curative Home with biodynamic milk and other produce. The first wholly biodynamic farm in the UK was Sleights Farm at Huby near Leeds, owned by Maurice Woods and converted in 1929.

In 1936 Dr Eugen Kolisko and Lili Kolisko came to settle in Gloucestershire and continued the scientific work they had been doing as co-workers of Rudolf Steiner.

In 1935 a separate Bio-Dynamic Association came into being, founded by Lady McKinnon and others. In 1944, David Clement managed to bring the two associations together and the present Bio-Dynamic Agricultural Association was formed, with David Clement holding the office of chairman until 1989. The association is still housed in Clent, in the West Midlands, now in the grounds of Sunfield Children's Home. There are approximately six hundred members, mostly home gardeners. There are several active regional groups around the country who meet regularly to study, take farm walks and make the preparations together. Preparations, books and other products are distributed from the main office in Clent by the trading company, B.D. Supplies. Conferences and workshops are organized from time to time, and twice yearly the association publishes issues of the journal *Star and Furrow* and a newsletter.

Biodynamics in the United States of America dates back to the early 1920s. The first person to make and use the biodynamic preparations was Henry Hagens, in Princeton, New Jersey, in 1925. The following year Charlotte Parker bought a farm near Spring Valley, New York, in order to grow quality vegetables for a restaurant in New York City. Two friends — Elise Stolting (Courtney) and Gladys Barnett (Hahn) — set off to learn biodynamic practices on the Keyserlingk estate in Germany, returning to introduce these methods in 1928.

In 1933 the first conference was held, bringing Ehrenfried Pfeiffer to the United States for the first of many visits. The Biodynamic Farming and Gardening Association was formed in 1938. Soon there were discussions about a winter school, a biodynamic newsletter, local meetings and centres for advice and supply of preparations. The biodynamic trademark was first registered in 1943.

The Association was based for some years at Kimberton Farms, Pennsylvania, under the direction of Pfeiffer who had moved to the States to live. After 1944 the headquarters moved to Chester, New York State, which was the home for many years of the Pfeiffer family. Ehrenfried Pfeiffer died in 1961, after a life spent both in active farming and in later years, dedicated to research and advisory work.

The Association headquarters is now back in Kimberton Hills, Pennsylvania, where there is an active farm and training

courses are organized. The magazine *Biodynamics,* founded in
the 1940s, is published regularly.

In Canada there are two biodynamic associations, the Society
for Biodynamic Farming and Gardening in Ontario, and in
Vancouver the Biodynamic Agricultural Society of British
Columbia, where some individual farmers and gardeners have
been using biodynamic practices since the 1950s. Their association started in 1973 and is active in holding meetings and
conferences, as well as supporting research and distributing
literature. They publish a journal, *The Stirring Stick.*

The Biodynamic Agricultural Association of Australia, founded
in the mid-fifties, is concerned with the redemption and enliven-
ing of dead soils, as well as training in farming skills. The association was established by Alex Podolinsky who has provided a
service of instruction and initial supervision for those wishing to
convert to biodynamics in Australia. A wide-ranging advisory
service has developed for this purpose, organized on self-help
lines. The Demeter symbol has been registered as a trade-mark
in Australia since 1967. Since 1981, the Biodynamic Marketing
Company (a non-profit-making organization) is the main dis-
tributor of certified Demeter produce to wholesalers and retail-
ers throughout Australia, and is also active in the ever-increasing
export trade.

In New Zealand, the first organized use of the biodynamic preparations took place in the early thirties. A small informal organization was formed in 1939 and the present association was
formed in 1945. It now has one thousand members.

    The association organizes the making of the preparations on
a large scale, and this is considered to be one of the reasons why
such consistent effects can be observed. According to their field
advisor, Peter Proctor, on almost all farms using the preparations, a deepening of the roots in grasses, an improved soil struc-

A Flowform in action

ture, greater clover nodulation and a mixing of layers can be observed after about eighteen months of biodynamics.

New Zealand farms vary in size from market gardens of less than a hectare to large pastoral holdings of one thousand hectares or more, producing sheep and cattle solely for export. On these large holdings, the farmers make great use of aircraft, helicopters and motorbikes. The association has adapted these methods to enable farmers to apply the biodynamic preparations over large areas. To stir the preparations on such a large scale, farmers often use Virbela Flowforms, which are meandering flow systems designed for water treatment, and a number of other applications.[6]

New Zealand soils have evolved in the absence of ruminants, as all native grazing creatures were birds. As a result, soils are

A biodynamic farm shop

often deficient in certain minerals and respond very favourably to the application of biodynamic preparations. Of the three hundred biodynamic farms, about fifty have attained the Demeter symbol for their produce, although there are still problems to be overcome with distribution and marketing, with consumer demand being as yet rather poor.

Biodynamics have been practised in South Africa since 1937 when Karl Adler, an Austrian emigrated there. The present Biodynamic Association of South Africa was founded in 1984 with activities mainly centred around Johannesburg and Cape Town.

Producers and consumers

Obviously there are differences in the development and growth of biodynamic farming in various countries in Europe, but

almost everywhere there is a connection between the demand
for products and the expansion of the number of farms.

In Germany and Switzerland some biodynamic farms have
their own local customers who come, for example, once or
twice a week for their vegetables, dairy products and (sometimes) bread. In France, vegetables and dairy products are
often sold in the organic section of local markets.

However, as soon as the relationship between the producer
and the consumer is no longer so direct, there is a need for a
guarantee and a brand name. The Demeter brand name was
used early on for biodynamic products, so that the origin of
the products could be identified by the consumers. The modern aim of 'chain protection' whereby the product is controlled
from the primary producer, via the processor and wholesaler,
as far as the shop, has thus been a practical reality in the
Demeter organization for a long time. Demeter products are
mainly available from so-called 'wholefood' shops. These
shops answer specific consumer demands, and not infre-
quently they are established as a result of consumer activities.
In some countries, including Britain, sales are higher through
some of the supermarket chains. In order to be sure of a regular and constant supply, there are also imports from other
countries.

Biodynamic methods have also been implemented successfully and quite extensively in the Scandinavian countries,
especially Sweden, with 147 biodynamic farms and market
gardens, Norway with twenty-nine and Denmark with sixty-
five. The entire area of Darlana, in west Sweden, is being
encouraged to join consumer co-operatives to enable all farmers to convert to biodynamic and organic methods. The
Swedish government is giving three-year grants to converting
or existing biodynamic and organic holdings nationwide.

This is the situation in northwest Europe. In Eastern Europe
no similar initiatives were possible until recently. Now there
are groups which have started to use biodynamic farming

methods in Poland, the Czech and Slovak Republics, Romania, Hungary, Estonia and Russia. These groups are active in the growing exchange of ideas and methods between Eastern and Western Europe.

In the Mediterranean area there has been a different development. The interest of consumers in the quality of agricultural products has come about only very gradually. It was particularly the demand of consumers further north for biodynamic rice and oranges and lemons, which served as an initial impetus. For example, a highly motivated and well co-ordinated organization of biodynamic farmers has been expanding rapidly in Sicily. This was initially with information provided by, and the supervision of, the German Demeter organization, and now has the help of an Italian information association and an Italian Demeter organization. Similar developments are taking place in Spain and Portugal.

Organic agriculture is relatively common in the United States of America. However, consumer demand is growing very slowly and this means that the distribution and sale of the products is often very difficult. Nevertheless, wheat and other arable products are exported to Europe. In America biodynamic farming is only a very small movement though it is active in several states where there are regional departments of the Biodynamic Farming and Gardening Association.

In other parts of the world the impetus for biodynamic farming is often centred on anthroposophical initiatives such as a Steiner school, or a residential home for the mentally handicapped where there is a call for biodynamic food.

In the developing countries the problems are completely different again. In those countries the important thing initially is not so much the biodynamic quality of the food, but simply to produce enough food. Then there is the enormous problem of drought and erosion which is common to many African countries. In fact there is as yet no properly organized system of biological agriculture and production. In many African countries it is a matter of working with farmers in such a way that they develop an understanding of, and pay attention to, the soil, humus and trees. In Kenya, for example, a successful development project has been set up by teaching farmers the art of composting on their farms. The manure from their cattle, household waste and all other plant and animal waste is properly composted. There is no shortage of manpower and the results are almost always convincing.

A biodynamic coffee plantation has been operating in Mexico for decades, and exports Demeter coffee to Europe. The local coffee farmers are starting to follow its example and are now using biological methods.

There is regular contact and discussion between all these associations and initiatives throughout the world, for example, at the annual international conference in Dornach, Switzerland, where anthroposophical insights are explored, and attention is given to practical problems and the exchange of experiences.

Research

Ever since the very earliest days of biodynamic farming, there has been close co-operation between research and practical activities. The farmers in the Experimental Circle and scientific researchers at the Goetheanum in Dornach worked together to develop biodynamic farming. Wherever possible, this co-operation still continues.

Research is carried out not only at the Goetheanum, but also in several areas of Germany as well as in Sweden and Denmark, the Netherlands, America, Brazil and Britain. In general, this research is carried out at small institutions without government support although there have been long-term comparative trial experiments carried out in collaboration with government bodies in Germany and Sweden. Some of these research areas will be examined further in Chapter 6.

The Goetheanum in Dornach, Switzerland

The scientists often adopt a phenomenological approach, which is concerned not so much with the physical analysis of a plant, as with the phenomenon of its whole external development. For example, the research can be aimed mainly at becoming familiar with a crop in its various forms during the development from seed to the fruit, and back to the seed again. This is often done together with the farmer or market-gardener who want to know more about the crops in their own soil, using their own manure to achieve better methods of cultivation and improved quality.

Other researchers are not so concerned with practical applications and work in areas of pure phenomenology, for instance, examining the laws governing the growth of the plant or the influence of factors such as climate and/or the zodiacal constellations. In addition, comparative research is carried out to demonstrate the effects of biodynamic preparations on plants or the effects of different types and quantities of manure. An initial problem which arose in research into qualitative characteristics is that there were only quantitative methods available to measure and compare these differences. Sensitive methods have since been developed for qualitative analysis, and these will be discussed further in Chapter 6. Finally, work is carried out in seed production, and in particular to develop and select varieties which are suitable for biodynamic methods of cultivation and which have a high nutritional value.

Advisory services

In modern agriculture it is often useful and practical for the activities of research, consultancy and agricultural education to be closely linked. But as research and consultancy services are generally based on current materialist views about the physical composition of plants and the soil, and pursue the general aim of high yields using chemical and technical labour-saving means, these services are of little value to biodynamic farmers except in some, purely technical, matters.

Thus it was clear that advice for farmers should be passed on by farmers with their own biodynamic experience, a system which can work very well. In Britain, there is a network of biodynamic farmers and growers who offer advice to newcomers in their own geographical area. The network is co-ordinated by a fieldsman who tours the country. This is the system used in a number of other countries, too, including Australia, New Zealand, Canada and the USA. Professional fieldsmen have also in recent years assisted farmers who were interested in switching from conventional to biodynamic methods — a considerable undertaking — in almost every country where biodynamic associations are established.

During the fifties and sixties, changing over to biodynamic methods was not such an enormous task because most farms at that time still managed dairy herds as well as arable farming, and therefore used their own cattle manure as well as artificial fertilizers. Nowadays this is no longer the case; farms tend to specialize and often they are run so intensively for high yields, that financially and economically it is hardly possible to switch back.

Thus the first job of an advisor, once he has informed the farmer about these methods of farming, is to find out, together with the farmer, whether the farm will be able to comply with the conditions and guidelines. If this question can be answered affirmatively, he can see what will have to change, and a schedule is drawn up for the implementation of these changes.

Nowadays the results of biodynamic farming are positively assessed by the governments of many different countries, firstly, because no herbicides and pesticides are used and much less manure is produced, and secondly, because the yields of so-called 'surplus' crops and milk are lower. As a consequence a number of countries subsidize the biodynamic information service or the advisors are employed by the government. This is the case in parts of Germany and especially in the Netherlands, for example. There the Biological Agricultural Team is part of the Agricultural Information Service and includes both biodynamic advisors and experts for the Association for Ecological Agriculture. Half of the costs are paid for by the government and half by the associations themselves.

Education and study

In a movement such as the biodynamic movement, starting from a non-materialistic view of nature, agriculture and society in general, it is important to offer opportunities for learning and discussion through lecture courses, as well as practical training. Educational facilities are provided in different ways in different countries, depending on the local customs and the agricultural education available.

In Britain, there has been a full-time training in biodynamics, both theoretical and practical, for over twenty years. The course is run at Emerson College in East Sussex and is also orientated towards the needs of developing countries. A number of farmers and gardeners have also been offering practical training for some years. Since 1989 this has become a two-year apprenticeship scheme operating in thirteen different farms and market gardens, supported by a series of intensive block courses in theory, organized by the Biodynamic Agricultural Association.

In the Netherlands, since 1947 there has been a three-year course at a college for biodynamic agriculture and horticulture, with its own working farm for practical experiences. In 1960 it was officially recognized as a college of further education in agriculture and horticulture. In addition, the Kraaybeekerhof study centre was established in 1977 to provide courses and training in many aspects of agriculture, nature and nutrition.

In Germany, the usual training for a farmer or horticulturist is to work on a farm for an employer who is able to train his apprentices. Once a week, these apprentices go to school for one day, and after three years they take an examination. This method is also followed on biodynamic farms. A student works on the biodynamic farm, follows the usual training and learns more about the biodynamic background in a separate course during the winter. In the last few years, there have also been courses in which theory and practice are more closely related and both are aimed at biodynamic methods. A one-year fulltime theoretical training course is offered at the Dottenfelderhof, near Frankfurt, for people who have completed their practical apprenticeship.

In Sweden, there is a two-year training course for the Scandinavian countries, while a biodynamic course recognized by the State was recently introduced in France.

There is a research and training centre for biodynamic farming in Brazil, and a centre has now also been set up in New Zealand to organize courses in that part of the world. In the USA, training is carried out on farms and there are intensive block courses for theory at different locations around the country.

This is by no means a complete list of all the possibilities for studying this field, but it does show that there are centres throughout the world working on biodynamic farming. Furthermore, several agricultural colleges now organize lectures in Biological Agriculture; for example, in Wageningen, Holland, where biodynamic agriculture is also covered wherever possible. Altogether there are five university chairs throughout Europe for biodynamic or organic agriculture.

~ 4 ~

Practical biodynamic farming

A biodynamic farmer obviously does many things in the same way as his colleagues who use traditional organic methods. The fundamental differences reside mainly in the farmers' ideological starting point. Comparing the three types of farmer — mainstream, organic and biodynamic — one might state (in rather extreme terms) that:

— the *mainstream* farmer's aim is a high yield and a high profit, to be achieved by means of technological improvements;

— the *organic* farmer's aim is to work in a way that is environmentally friendly and ethical with regard to animals. They do not buy artificial fertilizers to add essential elements to the soil and therefore use some form of animal or green manure for that purpose;

— the *biodynamic* farmer works on the basis of an awareness or sense that every living being has a link with the spiritual cosmic world, and that it is the duty of every human being to guide the life of these beings in such a way that these links can take place undisturbed. Furthermore, they work on the basis of the view that the Earth is a living organism and that a farm itself is a living organism.

Obviously there are many transitional stages and individual

attitudes between these three types of farmer. A biodynamic farmer will often use methods that an organic farmer believes to be necessary, for example, using a broad system of crop rotation. The fact that in the choice of the crops for his system of crop rotation, a biodynamic farmer may have different considerations, is not always clear in practice. There are other important factors, such as the potential of the soil on the farm, the crops which will grow in this soil, and how much money a particular crop will earn. The profitability of a farm is always a precondition for being able to farm using biodynamic methods.

The terms 'farm-produced foods,' 'farm-produced manure' and 'soil animal husbandry' are often used. In fact, these terms are a concise way of formulating the aims and starting points of a biodynamic farm.

The catastrophic developments in modern agriculture and the attendant environmental consequences and decline in quality of the products, are entirely related to the fact that the principle of soil-related farming has been abandoned. This clearly also shows that the only real solution to this situation in agriculture is biological farming which is based on this principle. A farm should not have more cattle than it can itself support and should use the manure on its own fields and meadows. Conversely, farmers should not grow more crops than they can produce using their own fertility-builders (including animal and green manure and similar fertilizers). It is self-evident that in this respect farms can work together, exchanging manure or fodder.

The mixed farm

As stated earlier, a mixed farm is the ideal type of biodynamic farm. It is not always possible to achieve this totally, but this is the aim; it is also mentioned in the guidelines as a requirement which must be achieved to some extent in order to be eligible for

*A biodynamic
polytunnel*

the Demeter trademark. In Chapter 2 we have already described
the cyclical activities of the farm, the cycle from feedstuffs to livestock, to manure, to the soil, which in turn produces fodder and
other crops. The way in which manure is treated is very important. It should be slightly rotted down before it is placed on the
soil so that organisms living in the soil are able to assimilate it
without any disruptive effect.

Because of the practical necessity to employ a broad crop
rotation, it is immediately clear that a farm with livestock will
have far more possibilities of achieving this. On this sort of farm,
fields are sown alternately with a mixture of grasses and clover
which is used as animal feed. A great deal of roughage can be
grown which the cows like to eat and this also enriches the soil

Table 1. Objectives in biodynamic and conventional farming

BIODYNAMIC OBJECTIVES	CONVENTIONAL OBJECTIVES
A. *Organization*	
Ecological orientation, sound economy, efficient labour input	Economical orientation, mechanization, minimizing labour input
Diversification, balanced combination of enterprises	Specialization, disproportionate development of enterprises
Best possible self-sufficiency regarding manures and feed	Self-sufficiency is no objective; importation of fertilizer & feed
Programme dictated by market demands	Stability due to diversification
B. *Production*	
Cycle of nutrients within the farm	Supplementing nutrients
Predominantly farm-produced manuring materials	Predominantly or exclusively bought-in fertilizers
Slowly soluble minerals if needed	Soluble fertilizers and lime
Weed control by crop rotation, cultivation, thermal	Weed control by herbicides (cropping, cultivation, thermal)
Pest control based on homeostatis and inoffensive substances	Pest control mainly based on biocides
Mainly home-produced feed	Much or all feed bought in
Feeding and housing of livestock for production and health	Animal husbandry mainly orientated toward production
New seed as needed	Frequently new seed

BIODYNAMIC OBJECTIVES	CONVENTIONAL OBJECTIVES
C. Modes of influencing life processes	
Production is integrated into environment, building healthy landscapes; attention is given to rhythms	Emancipation of enterprises from their environment by chemical and technical manipulation
Stimulating and regulating complex life processes by biodynamic preparations for soils, plants, manures	No equivalent to biodynamic preparations; use of hormones, antibiotics, etc.
Balanced conditions for plants and animals, few deficiencies need to be corrected	Excessive fertilizing and feeding, correcting deficiencies
D. Social implication, human values	
National economy: optimum input/output ratio regarding materials and energy	National economy: poor input/output ratio regarding materials and energy
Private economy: stable monetary results	Private economy: high risk, gains at times
No pollution	Considerable pollution worldwide
Maximum conservation of soils, water quality, wild life	Using up soil fertility, often erosion, losses in water quality and wild life
Regionalized mixed production, more transparent consumer-producer relationship; nutritional quality	Local and regional specialization, more anonymous consumer-producer relationship; interested in grading standards
Holistic approach, unity between world conception and motivation	Reductionist picture of nature, emancipated, mainly economic motivation

Table 2. Farming costs and yields: conventional and biodynamic (based on the annual report of the ministry's accounting service, Sattler, Pers. Com.)

Costs and yields	Talhof farm	Comparable conventional farms in the district
Expenses for fertilizers or materials for preparations and straw (€/ha/year)	3.95	75.15
Yields: grains (kg/ha/year) milk (kg/cow/year)	3600 4399	2900 3376
Bought-in concentrate (€/cow/year)	17.90	115.00
Hectares per worker	10.80	9.70
Income per hectare (€)	920.00	568.00
Income per worker per year (€)	9585.00	5500.00

with large quantities of organic material and with nitrogen derived from the leguminous crops. The farmer can add the preparations to the manure and compost it in good time. In his barns he will store straw from his own cereals.

Calculations are drawn up nowadays in so-called 'mineral balance sheets' to find out how many nitrogen compounds such as ammonia and nitrates are used, and how many are lost on a farm, thus polluting the water and the air. Such audits show that these proportions are most favourable on a mixed biodynamic farm.

Manure and compost handling

Present ideas on fertilizers are particularly concerned with the application of the minerals, nitrogen (N), phosphorus (P), and potassium (K). These can be applied in mineral form as artificial fertilizers, or in the form of animal manure and compost.

This view is certainly supported by many organic farmers. They believe that it does not make any difference whether slurry is used, or chicken manure, or pig manure, as long as the right quantities of N, P and K are added to the soil. This way of thinking is also familiar to biodynamic farmers and gardeners and it cannot be denied that the quantities of N, P and K which are applied are significant. However, practice and experience on biodynamic farms have shown that the forms in which these substances are administered is at least as important.

Another way of formulating the aim of fertilizing the soil biodynamically is to say that we fertilize the living fertile soil,

Making compost

A compost heap

and not the plants. The soil may benefit from minerals, but it certainly also benefits from organic matter and life forces, and even the influences of the planets and stars.

The soil in itself should also be considered as a living organism, with its own equilibrium and processes, its respiratory and digestive systems. Therefore the form in which the manure is applied certainly has an influence. The organism of the soil as a whole must be able to assimilate the manure and derive nutrition from it. Therefore the best form is manure which has been rotted down in advance; in other words, manure in a composted form.

A field plot experiment was carried out in the Biodynamic Institute in Järna, Sweden, over a period of forty years. The following table was compiled with the results gathered over the first eighteen years, and illustrates the effects of different types of manuring.

Treatment	Full bio-dynamic	Crude manure	½ manure, ½ NPK	Control	Fertilizer NPK	NPK
Fertilizer: kg N/ha/year	82	93	61	0	56	11
Yields, expressed as t/ha cereals	4.86	4.90	5.03	3.77	4.83	4.87
Bulk density of soil						
top layer	1.14	1.09	1.10	1.10	1.14	1.16
subsoil	1.33	1.29	1.42	1.50	1.51	1.48
Organic matter (total N)						
topsoil %N	0.24	0.24	0.25	0.25	0.26	0.26
subsoil %N	0.14	0.17	0.08	0.16	0.12	0.09
Earthworm burrows > 1.5mm/m^2	100	111	53	22	11	16
mg CO_2/100g of soil	125	108	91	83	75	81
Dehydrogenase, TPF/10g soil	547	377	302	213	211	258

Table 3. Soil characteristics as influenced by different systems of manuring over 18 years (Pettersson & Wistinghausen 1979)

The crop rotation included wheat, potatoes, vegetables and clover-grass. The figures show an improvement of biological factors under organic and biodynamic soil treatment. What can be seen clearly in the biodynamically treated plot is the improvement to the organic content and bulk density of the subsoil. So a much greater depth of soil becomes available for the plant roots, and earthworms are aerating the soil at a deeper level.

On the mixed biodynamic farm, manure and the vegetable waste matter are placed together in a compost heap, where they rot down. These are also life processes which can only take place properly if the life forces in them interrelate in the correct manner. This

Compost heaps

means that we should make sure that the different elements — the earth element (that is, minerals and particles of soil), the water, air and warmth elements — are all present in the correct proportions. Therefore, the compost heap should not be too wet, but it should not be too dry either, and if the proportions are right the heating up process will develop automatically, as the compost heap starts to work. Sometimes some ground basalt is added to the compost heap, and when there is vegetable waste, some lime is also sprinkled on. Depending on the crop to be grown, this compost can be spread on the fields or garden in a lightly rotted-down form, or when it has rotted down completely and is almost like soil.

When the compost heap rots down, this is a digestive process. The vestiges of plants, straw and animal manure all lose their form, colour and properties. These were not incidental, but developed in living animals and plants according to particular patterns, and on the basis of particular astral and etheric influences. These forces are once again released during

Burying a stag's bladder
Biodynamic compost preparations of camomile sausages

Preparation 506

the digestive process. They relinquish their material form and are partly used by bacteria, moulds, springtails, earthworms and many other lower animals, each converting the matter in their own way as they live and eat. However, they are also used again to create new substances and humus in the compost heap. They are available, as it were, for all those organisms that live in and on the earth, when the manure is used.

The biodynamic compost preparations

The decomposition in the compost heap and its efficacy when spread over the land can be greatly enhanced by the addition of six compost preparations. They are produced from six plant substances:

Preparation 502   yarrow flowers
Preparation 503   camomile flowers
Preparation 504   stinging nettle (the whole plant above
                  the root, in full bloom)
Preparation 505   oak bark
Preparation 506   dandelion flowers
Preparation 507   valerian flowers

Various preparations

These preparations are made up in various ways, using specific animal organs. Like the spray preparations (described on p. 81) they are made by some farmers themselves once a year. Or they may be made by a group of farmers or gardeners in the same region. They can also be purchased from the biodynamic association in each country. This is often the most practical solution for the home gardener.

On a garden scale, for a compost heap measuring about three cubic metres, a pinch (taken between the thumb and index finger) of each of these substances is placed in holes made in the compost heap with a stick. Approximately 1 cc of the valerian juice is stirred into three litres of water and then sprinkled over the compost heap with a watering can with a rose. On larger heaps, such as on a farm, more sets of preparations are applied at three metre intervals.

In this way an almost living organism develops once again. It has digestive and respiratory processes, develops its own

heat and has a complex of life forces and astral forces in which the substances fulfil the function of monitoring and regulating organs.

In experiments carried out in cooperation with an agricultural school and government research station in Germany, the preparations 502–507 were shown to increase the exchange capacity of organic matter in the finished compost heaps. They have also been seen to even out fluctuations in temperature in the compost heap to some extent.[7] Several effects have been shown in experiments using slurry treated with the compost preparations, such as a reduction in the losses in nitrogen.[8]

Animal husbandry

An old picture of a farm where a pig is rooting in the mud, chickens are wandering around pecking at grains or looking for earthworms in the manure, and cows are dreamily grazing in the meadow, is a beautiful image of the typical behavioural instincts of animals. When they are able to behave instinc-

tively, they show very clearly what they need and what they like to do. Farmers have observed these patterns for centuries, and usually ensure that their animals are able to live in such a way as to fulfil normal behaviour.

However, in modern 'industrial' farming such considerations are no longer practical. Efficiency and profitability require as many labour-saving devices as possible, because employee wages are high. It follows that as few hours as possible should be spent actually looking after the animals. The ensuing ways in which many animals are kept on modern labour-saving farms have given rise to strong ethical objections. The hellish sight of overcrowded and confined battery hens with their beaks burnt off (to prevent them from pecking one another) is certainly far removed from the old image of the chickens roaming and scratching happily in the farmyard.

Chickens

Chickens on a biodynamic farm have a spacious covered area where they range freely and also have an outside run. They are fed with

grain and corn and they peck these themselves, making a great deal of noise and scratching the ground with their feet and beaks, which are not burnt. They are also given fodder and meal ground from raw materials which have been cultivated using biodynamic methods. The difference between organic and biodynamic methods for chickens is not very great. On the basis of the ethical views which apply in both organic and biodynamic farming, the chicken runs are chicken-friendly so that the chickens can live according to their own nature. The differences between the two methods are mainly concerned with the specific production of chicken feed.

Cattle

Of all the domestic animals on a farm the cow is by far the most important, providing milk, calves, and meat, as well as leather. Cattle were also used for drawing wagons and this is still the case in many countries. Another extremely important product is cow manure, neglected in modern agriculture as superfluous but valued highly on biodynamic farms.

Cows are grazers, *par excellence.* They have an elaborate digestive system, which they use to assimilate large quantities of grass and other roughage, and convert them to milk. A grazing herd or a herd lying down chewing the cud is a beautiful image of the harmonious relationship between the soil, plants and animals. Cows are able to assimilate so much vegetable material during the lengthy and thorough process of chewing the cud, that they can produce an abundance of milk, as well as the manure which feeds the soil.

The biodynamic farmer knows that a cow is made to process a great deal of roughage — grass, hay and silage — and he will rear his cows in such a way that this quality develops properly. He will give very few concentrates to the cows and sometimes even tries to do without concentrates (that is, meal) altogether, as cows do not naturally eat any grains or pulses.

In modern cattle-farming, large amounts of (imported) grain are fed to the cows to achieve a high milk yield. Cows which produce 10,000 litres per year are no longer exceptional.

Milking

In contrast, the aim of a biodynamic farmer is to ensure that the cows are healthy, have a long lifespan and a good, regular milk yield and sound calves.

On larger biodynamic farms the farmer often keeps his own bull. This also corresponds with the aim of running the farm as an organism in which the farm has its own herd and where the soil, the fodder and the animals interrelate as they should. On some biodynamic farms, the cows are inseminated artificially, as they would be on conventional modern farms and on most organic farms.

The young calves are fed on cow's milk at first, and after several weeks this is supplemented with carrots and hay. They must have access to fresh air and movement.

It is also important that mature cows have plenty of light, air and movement. There are quite a few biodynamic farms with loose housing in which the animals walk around freely or lie on a thick layer of straw which absorbs all the manure and urine. This produces large quantities of stable manure and gives the cows a warm bed to lie on, as the deep litter manure tends to generate heat. The air is fresh, as the ammonia and other substances are absorbed by the straw.

These barns are quite different from the common barns with cubicles on modern farms where animals can also roam around freely or lie down in a cubicle, but always in the rather acrid smell of slurry. The floor of these cubicles has slats so that the manure and urine disappear into a large cellar. Therefore there is no 'stable manure' formed with the straw, but a thin slurry. The bacteria which develop in this liquid slurry are quite different from those which thrive in stable manure, which is rich in straw, and therefore oxygen. The bacteria in slurry assimilate oxygen from the manure giving rise to compounds with a low oxygen content; this has a negative effect on the life in the soil when the slurry is spread on the fields (after a thick application of slurry, the earthworms cannot do their job properly.) The stench of slurry is also quite different from the smell of stable manure. However, it is possible to improve the quality of slurry by aerating it and by the addition of chopped straw, crushed basalt and use of the biodynamic preparations. Slurry treated with the preparations has been shown to increase the length of roots and their dry-weight, when used in growth experiments on wheat.[9]

However, there is not a deep litter barn on every biodynamic farm. A farmer who is converting to new methods cannot simply build a new barn just like that. The multiple barn, sometimes with an automatic mucking out system, in which the cows are tethered and lie on straw, is also common. In this case the manure and the urine are kept separately.

A special characteristic of biodynamic farms is that the cows are not de-horned. The habit of removing cows' horns arose when they were kept in cubicles in a type of barn designed to save the farmer the labour-intensive job of mucking out. Cows belong in a herd in which every cow has its own place, and the cows which are higher up in this order sometimes like to remind the others of their position by nudging them with their head. Therefore, a cow which lies down in a cubicle where another cow higher up in the order wants to lie down, is in danger of being butted if it cannot move quickly enough. When the soft part of the udder is nudged this can cause nasty wounds. This is why the horns are sawn off or why the horns of young cows are prevented from growing by the use of chemicals or by burning them.

The biodynamic farmer's view is that the horns of a cow — like the hooves — have a function and are important both for the animal's welfare and for the metabolic and lengthy digestion processes.

In the cow, digestion is of paramount importance. She grazes and chews the cud just like sheep, goats, deer, antelopes, buffaloes and so on. All these ruminants have a strongly developed stomach sub-divided into four stomachs (the rumen, the reticulum, the omasum and the abomasum) so that it is possible to digest thoroughly all the food that is eaten, mainly very rough and fibrous material. However, there are differences between the stomachs of the different species and these differences between their stomachs are related to differences between teeth and the shapes of antlers and horns.

In other words, there are important differences between the essential characteristics of these animals, expressed in their teeth, horns, behaviour, and the quality of the manure they produce, all of which have a distinct influence on the organism of the farm. The cow, with its dreamy, grazing existence, its chewing of the cud, its manure, and its horns which are like a sort of counterpart to the heavy and elaborate digestive system, is best suited to the organism of the mixed farm as it exists in Europe.

The organism of the cow is therefore not complete without its horns, and certain qualities are diminished when the animals have their horns removed. As mentioned above, there is also a connection between the cow's horns and its hooves. It is

interesting to note that where cows have been de-horned, there are often problems with the animals' hooves. They become inflamed, the cows cannot stand or walk properly and have to be destroyed.

The horns of the cow are also very valuable on a biodynamic farm because they are used for the manufacture of the manure preparation and quartz preparation, which will be described below.

Pigs

When you examine the essential characteristics of pigs, you soon come to the conclusion that they are highly vital animals. You would almost say that they have a great enthusiasm for life and a lively interest in their environment, which they relate to very actively. They do this in groups, examining, sucking and tasting everything with their rooting, sniffing noses, greedily biting. Their vitality is also evident from the fact that they have only a four month gestation period and farrow seven to ten piglets at the same time. These piglets then reach sexual maturity nine months later.

In the current commercial pig-rearing industry this vitality has been exploited in a terrible and unacceptable way. It appears that pigs are able to tolerate a great deal as regards prison-like accommodation, industrially processed foods and the administration of hormones, without being finished off altogether.

Young weaners, instinctive rooters which are particularly active and inquisitive when they are young, are kept in concrete cells on a slatted floor, imprisoned and with no room to move. They become so frustrated that they start biting and tearing at each other's ears and tails, and can become unable to stop sucking and biting.

On a biodynamic farm pigs are given straw in their pen and they have a run to move about in. At a very young age they go out into the field. A deep litter pen with lots of straw and fodder, and occasionally a few spadefuls of compost, will also meet their requirements.

Pigs are omnivorous. In fact, they will eat all the surplus on the farm, tidying it up. Anything that is not good enough for human consumption can be eaten by pigs and thus they do very well, for example, on a farm where cheese is made and the whey is given to the pigs. Inferior grain, or potatoes which are too small, can also be given to pigs.[10]

The sick animal

Brief mention should be made here of treatments for sick animals. If animals are cared for in a way which allows them to display their natural characteristics, if they are fed from fodder grown on the farm, given suitable housing and exercise, then the incidence of illness is dramatically reduced. This has been borne out by biodynamic farmers the world over. Another important factor is breeding for health and longevity rather than high performance and high production.

When an animal does fall ill, sometimes it is necessary to call in the vet immediately. But often the person in charge of the animal can take measures before the illness becomes acute. A whole range of anthroposophical veterinary medicines have been developed for this purpose but sadly are not available in all countries.[11]

Arable farming and vegetable cultivation

Obviously the fertility of the soil is the most important aspect for the cultivation of plants, particularly as it is not possible to add a little artificial fertilizer when you are cultivating vegetables or arable farming using biodynamic methods, and diseases cannot be dealt with simply by using herbicides and pesticides.

Biodynamic farmers devote a great deal of attention to the structure of the soil, the level of fertility and the life in the soil. They do this by means of:

— the way in which they cultivate the soil
— crop rotation
— fertilizing with composted manure
— fertilizing with green manure
— spraying preparations made from manure and quartz concentrates.

In addition, a great deal of attention is devoted to controlling the growth of weeds.

The biodynamic spray preparations

There are two types of preparation: the compost preparations described earlier, and the spray preparations. These are known as preparation 500, made with cow horn and cow dung, and preparation 501, made with cow horn and ground quartz. As is

Cabbages
Leeks

the case with the compost preparations, the numbers merely denote a catalogue listing when they were first made, and have no significance.

Preparation 500 is exposed to environmental influences in the ground for one winter, then dug up and stored. When being prepared for spraying, a small amount is stirred thoroughly in water for an hour. On a garden scale, 50 g of the preparation 500 can be stirred in ten litres of water; this can then be sprayed on approximately 2,500 m^2 of land. On a farm scale, 40–60 litres of water and 250–300 g of horn manure will be required per hectare. For this scale, the stirring is done in barrels and the spraying from the tractor. For smaller areas, a knapsack sprayer is used.

The manure preparation particularly affects the biochemical processes in the soil, in that part of the soil where water, earth and humus interact. This helps the plant to take firm root and interact with the life of the soil.

Preparation 501 is subjected to environmental influences below ground for one summer. Even smaller amounts of this preparation are needed. 5g of preparation 501, stirred for an hour in 60 litres of water, is adequate to spray one hectare of field crops. A slightly higher concentration is used on a garden scale. It is sprayed, using a very fine nozzle attachment, on to the green leaves of plants during the growing season. It can be sprayed on a crop during several different stages of its growth. In young plants, it encourages the process of assimilation and the development of a strong structure. In more mature plants, it aids the maturation process as well as the plant's fragrance and flavour which in turn will enhance keeping qualities.

It has been shown in experiments carried out in Germany, America and Sweden, that the spray preparations have several discernible effects. These include: increased yield of a number of crops; increase in quality (such characteristics as the percentage of true protein, nitrate and soluble amino acids, sugar and other carbohydrates, and vitamins, as well as the activity

Preparation 500

of various enzymes, can be changed by these preparations); and improved keeping quality.[12] For experimental data, see Chapter 6.

Soil cultivation

There are many different tools for cultivating the soil. These are used with a tractor to plough, hoe, roll, rake, harrow and cultivate the soil. In this respect, the organic farmer uses the same tools as conventional modern farmers, though, as far as possible, he will try to avoid spoiling the structure of wet ground with heavy machinery.

In fact, experience shows that the better structure of the soil on biodynamic farms means that lighter tractors can be employed, using less horse power.

Crop rotation is a very important routine for the health and fertility of the soil. There should be a rotation, for example,

*Stirring the
preparations*

between crops with deep roots and those with surface roots; crops which leave large root-stocks behind, such as cereals, and crops which leave very little behind, such as potatoes; crops with which weeds can easily be hoed away or chopped down, and those where this is not so easy. Therefore there are many factors which a farmer should consider when drawing up plans for crop rotation.

The soil should be fertilized with composted stable manure, that is, manure which has partly rotted away. This can be spread with a muck spreader, ten to thirty tons being spread per hectare. Green manure crops are also often used as a fertilizer in organic farming. This means that a crop is sown and not harvested, but is dug into the soil to enrich it. Clover and other leguminous plants are often used as green manure because they are able to absorb the nitrogen from the atmosphere, and thus add it to the soil.

When the soil is tilled and prepared for sowing, preparation

Sweet oranges

500 should be sprayed to stimulate all the life processes, transformation processes, germination and rooting of the plants. At a later stage, preparation 501 is sprayed on the green plants.

Then there is the matter of weeding. For conventional farmers contemplating converting to other methods, this is always the main question. How will they cope with the weeds without using chemical weed killers?

Fortunately, now that there is general resistance to the use of herbicides, better machines have been developed in recent years for weeding mechanically. It is a matter of hoeing, harrowing or earthing up the soil at the right moment with machines adapted to the particular soil and crop concerned. In this respect there have been many developments carried out by farmers using organic methods, and these are gradually also being used in combination with ordinary modern farming methods.

Fruit cultivation

The biodynamic cultivation of fruit is still in its early stages. Farmers specializing in this area are real pioneers. However, the same can be said, in varying degrees, of all farmers and market-gardeners who use biodynamic methods. Without a certain pioneering spirit, it is best not to start.

The modern conventional cultivation of fruit makes very intensive use of chemicals. There are chemicals for fertilizing, for thinning out the fruit, for encouraging the right colour, and of course for combating moulds and insects and killing weeds. Varieties of apples and pears grown nowadays are all very sensitive to disease and therefore new varieties are being developed which are more suitable for biodynamic methods of

Apples

cultivation. Ideally these should not only be resistant to disease but taste good, keep for a long time and have a good colour.

In biodynamic cultivation, composted manure is used for fertilizing. Another measure which is often used is to spray the trees in winter with tree-paste, a mixture of clay, manure and lime, as a fertilizer for the trunk and branches. This also has a strengthening effect and prevents diseases.[13]

Beekeeping

Ideally, every biodynamic farm or garden should have its own bee-hives. There are also beekeepers who work using biodynamic methods. Bees are creatures which live mainly in light, air and heat and they are an essential part of the interrelationship of natural forces.

A colony of bees is in itself an organism. A hive full of bees, thousands altogether which belong to a single colony with one queen bee, functions as a highly co-ordinated animal organism. The colony has its own inner world, its 'astral body' with its own heat, and a rhythm and reproductive cycle controlled by the queen bee. Bees live entirely in the world of warmth, light and flowers from which they take nectar and pollen. They hardly touch the earth and drink only a little water. They are true creatures of the sun and play an important role in the organism of a farm or garden. Therefore it is not surprising that pure honey has countless healing qualities.

Modern beekeeping methods are strongly influenced by chemical and technological processes and drastic interference in the development of the colony. For the last ten years in Europe and the Mediterranean, and only recently in Britain, beekeepers have been plagued by the varroa mite, a minuscule spidery creature which lives on the body of a bee. If undetected in time, this pest can wipe out whole colonies, and the only available treatment to combat it, once discovered, is in chemical form.[14]

Maria Thun

Biodynamic beekeepers view the colony as a living unit which should only be interfered with to a limited extent, always using natural methods. For determining the best times for this intervention, reference is made to the *Biodynamic Sowing and Planting Calendar* by Maria Thun.

Rhythms of sowing and planting

Maria Thun lives in the state of Hessen in Germany and works to promote ideas on biodynamic farming. In her youth, she often came across traditional customs used by farmers which related to the lunar cycle. She wished to study whether these ideas were correct or incorrect, and for almost forty years she has researched into the influence of the moon on the life and growth of plants and animals. The influence of the waxing and

waning moon, the full and the new moon, has always been recognized in ancient tradition. To twentieth-century people, this
knowledge seemed old-fashioned and was no longer taken
seriously. Nevertheless, to take only one instance, it is a known
fact that more babies are born just before a full moon than
afterwards.

By carrying out lengthy and repeated tests with sowing, Maria
Thun established that there are lunar effects on the growth of
plants. The results of her tests were not related primarily to the
full moon and the new moon but to the orbit of the moon
through the signs of the zodiac. The moon rotates around the
Earth and, seen from the Earth, it is always in front of one of the
twelve signs of the zodiac. There is a different effect on growth
processes depending on which sign is behind it.

The tests were repeated many times and clearly showed that
each of the twelve signs of the zodiac has a special relation with
one of the four life forces which are manifest in the 'elements'
earth, water, light/air and heat. All four forces can be seen to have
an effect on the life body for the growth and development of the
plant. We also see that the earth element has a particular effect
on root formation, the water element on the formation of the
leaves, the light/air element on the flowers, and the heat element
on the formation of fruit and seed. This is significant for our vegetables, flowers and fruit.

All our cultivated varieties have been developed in such a way
that either the root is strongly developed (for example, in beet-
root and carrots) or the leaves (for example, in cabbage or let-
tuce), or the flowers or the fruit. These cultivated crops
concentrate on developing one particular part of the total plant,
and to do so they need extra life forces. The particular aspect of
development with which we are concerned here can be further
enhanced by choosing the days for sowing, planting and working
the soil in such a way that the moon appears in one of the signs
of the zodiac which influences that part of the plant.

This seems more complicated than it really is. The moon's

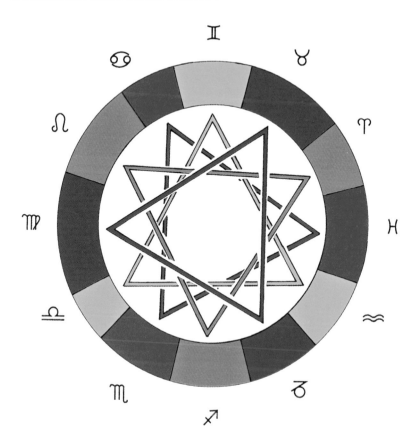

Zodiac diagram

The four trigons of the zodiac that act on root, leaf, flower and fruit, based on the work of Maria Thun.

orbit through the signs of the zodiac takes twenty-seven and a third days. Thus it is always in any given sign for a duration of two or three days. The diagram here shows how the influences of the signs of the zodiac alternate.

It takes about nine days for the moon to move from one water sign through the air, earth and fire signs, to the next water sign. Thus, if it is not possible to plant the crop on the right days because of the weather conditions or a lack of time, the most appropriate date to plant them will be nine days later. If this is too late, the plants can simply be planted on another day. These measures need to be approached in a flexible and pragmatic way.

Many farmers and market-gardeners make grateful use of the sowing calendar. This does not mean that they have much bigger yields, but many people believe that their produce is constant, with fewer diseases and of an excellent quality. Maria

Thun's suggestions have been taken up in several research projects.[15] Her own research results have been published regularly in the *Biodynamic Sowing and Planting Calendar.*

Calendars following the same system are drawn up every year in different countries.[16] They describe not only the moon's orbit through the signs of the zodiac, but also the positions of the other planets in relation to the Earth and the moon, which also influence growth and weather conditions.[17]

For the sake of clarity, it should be pointed out that the orbit of the moon around the Earth, as seen from its position in the signs of the zodiac (the sidereal lunar orbit) is shorter than its orbit in relation to the sun, because the Earth is moving around the sun at the same time. This is the synodic lunar orbit, which takes 29.5 days.

Beekeepers who work with biodynamic methods notice how sensitively colonies react to lunar influences. By choosing the right days to work with the colonies, they can be stimulated to concentrate on different activities to suit the beekeeper's plans.

The landscape and conservation

The concept of a farm as an organism in which all the component parts, like organs, look after the unit as a whole, allows room for the constituents of the landscape. Biodynamic farming methods are not opposed to operating on a large scale provided that the hedges, woods, streams, ponds and marshes are allowed a place in the whole. Because the effects and proportions of the life forces and astral forces in these organs, each with their own animal population, are different, they interrelate like a healthy, multi-faceted organism. The greater diversity of wild plants, birds, insects, reptiles or amphibians which are able to live in a varied landscape, is indicative of this. If, furthermore, there is no use made of agricultural toxins or excessive quantities of slurry, this means that some natural conservation is already taking place. In biodynamic farming no division is made between natural areas and intensive farm-land; rather they are integrated. In fact, our varied European landscapes are actually the result of the way in which farmers worked in past centuries. They cut down the woods and ploughed the fields, planted trees and hedges, dug ditches and built dams. This is how the cultivated landscape of Europe was created.

Social aspects

Of necessity, farmers and market-gardeners are very price- and cost-conscious. This is because in order to farm using modern methods, a great deal of investment is required in land, machinery and buildings. Prices are constantly under pressure from foreign competition. In fact, production in Europe is far too high, but the same fundamental problems apply throughout the world. Everywhere farmers aim at higher production to compensate to some extent for the low prices. As we know, this aim for ever higher yields has resulted in the use of bigger and bigger machines, more artificial fertilizers and more weed killers and pesticides, with the well-known unpleasant consequences for the environment. The damage caused by all this and the costs incurred are not charged to the farmers but are carried by society as a whole.

Therefore, the prices which farmers receive for their products do not cover the real costs. The costs which result from conventional methods of production — for example, in the form of higher treatment and purification costs for drinking water — are charged to the consumer. It therefore emerges that the low market prices of agricultural and horticultural products are an illusion. If all the real costs resulting from modern methods of farming and market-gardening were reflected in the price of the final product, this price would clearly be much higher.

In biodynamic farming and market-gardening the reverse situation applies. Because the farmers and market-gardeners use different methods on their land and for their manure, and do not use any chemicals, in particular no weedkillers, they do not pollute the soil, water or the air. On the other hand, their methods of fertilizing and dealing with weeds are more labour intensive and therefore more expensive, so that the cost price of their products is higher. This means that the consumer of bio-dynamic products pays twice for the prevention or reversal of environmental pollution. They pay once for the higher price of their organically cultivated food, and again through taxes and statutory water charges for water purification. It makes sense to prevent this environmental damage in the first place by means of organic, environmentally friendly methods.

The difference in price between conventional modern products and biodynamic/organic products also comes from the fact that yields are ten to thirty per cent lower. The transport and distribution system for the organic farming sector is on a smaller scale, too, and therefore more labour intensive than for normal modern distribution and processing methods.

Trade, sale and control

As discussed earlier, the development of biodynamic farming was originally actively promoted and encouraged both by producers and consumers. Farmers and consumers were in contact with each other and, through consultation, tried to arrive at acceptable prices and a realistic system of distribution. This co-operation led to the creation of distribution centres and, in due course, retail outlets. In future, large-scale distribution will become increasingly common, as well as the anonymity of consumers and producers in relation to one another.

If biodynamic and organic farming is to develop any further, the products will have to be sold in large quantities, preferably at a large number of sales outlets. Natural food shops are expanding to become small supermarkets of organically cultivated products. On the other hand, existing supermarket chains are increasingly revealing an interest in organic products, as their customers start to ask for them.

While such developments are welcome, great care should be taken to prevent the distinctive identity of organically and bio-dynamically grown products from being lost. To this end, joint efforts are being made to achieve quality control of organic products at a European level. For consumers it should be clear which products are genuinely 'organic' and which are labelled only for the sake of complying with the current 'green' trend.

The EC has even introduced a rule with which all the member countries are obliged to comply, to draw up identical basic norms and directives for recognized organic farming methods and organizations, including the processing of primary products to make cheese, bread and muesli and so on. If a producer wishes to market a product with the label 'organically cultivated,' they will be legally obliged to comply with these norms and be subjected to controls. These controls apply not only to the farmer but also to processors and distributors. Similar controls for production methods will also become compulsory for imports into the EC.

As the organizations involved have already carried out a great deal of work within the International Federation of Organic Agriculture Movements (IFOAM) to arrive at a consensus of basic norms, this international federation, which is active in various European countries, was able to prepare these rules in consultation with national officials and officials in Brussels, so that the broad lines corresponded with organic farming practices. Contact has also been established on this subject with IFOAM members in America and other continents.

Unfortunately, controlling the method of production from the farm via the processor to the importer and dealer requires a great deal of work, and is therefore expensive. These costs are added to the price of organic products which does not have a very favourable effect on their relative price.

The official control required by the EC amounts to a confirmation of organic farming methods and processing and will probably have an effect outside the EC, in America, Eastern

Biodynamic farming at Botton Hall, a Camphill community

Europe and beyond, because products are imported and exported to and from these countries. A development in the last two years is the attempt of dealers in Demeter products in Europe to establish a certain degree of policy agreement in the form of International Demeter Guidelines. In this way an attempt will be made to bridge the gap between the primary producers, the farmers, and the distributors. The Demeter traders do not wish to take over a market at the expense of the producers; rather they wish to create the possibility for biodynamic farming methods to survive.

The therapeutic effect of biodynamic farming

The starting precepts of biodynamic farming, concerned with caring for life processes and an understanding of living creatures, mean that this method of farming is eminently suitable for social and therapeutic institutions.[18] It has been widely applied in the first place in support of mental health care, in foundations based on anthroposophical ideas, such as the Camphill communities and villages in Britain. Working on a farm, looking after the cattle, the plants and the compost as well as preparing and enjoying the high quality food produced for the institution by the farm, all contribute to a therapeutic framework and daily routine. People who temporarily wish to find a new orientation in their lives, often find strength and support from the biodynamic farming practice on a therapeutic farm. Biodynamic farming methods are equally used in mainstream institutions.

In a number of Steiner Schools where sufficient land is available for a garden, lessons are given in the care of plants and the soil. There are also biodynamic farms which provide accommodation for school groups so that children can learn about farm activities and join in by helping. The children take turns caring for the animals, mucking out the stables, picking beans, digging up potatoes and helping with general work.

Land ownership

It is the farmers who work the land and are responsible for its fertility. But are they solely responsible, or should all people in a particular geographical area feel a common responsibility? This is a very real question to those who are concerned with biodynamic farming methods, both producers and consumers.

If soil is incorrectly worked, or exhausted or eroded, it can lose its fertility within a very short span of time, even within one person's lifetime. Signs of erosion appear particularly rapidly in hilly regions. In this way the potential for food production for our ever-growing population is destroyed.

Short-term profits do not apply to farming. It is always necessary to think in the long term. A farmer who is responsible for the management and cultivation of a particular area of

land should have the main aim of constantly improving his land so that it is better for his successors than when he started farming. Nevertheless, in view of the economic balance in modern times, every farmer must make a sufficient short-term profit to pay annual interest and capital repayments, or the rent. Good farmland is expensive and the investments required to establish a farm are very high. In various countries in Europe and America, people who are interested in biody-namic farming methods have tried to find other forms of land ownership and other forms of interrelationship with their clients. On the basis of the view that maintaining the fertility of the land should no longer be a private matter, the land could be owned, for example, by a trust. This trust would rent out the land to a farmer with an express clause in the contract that biodynamic farming methods must be used. In its articles of association, the trust would guarantee that land which is farmed in accordance with biodynamic farming methods will no longer be sold on the open market.

Obviously there are other conceivable forms of co-operation between landowners and those responsible for farming the land. For example, there are some cases in which a farmer or market-gardener collects a circle of interested people/con-sumers who undertake an obligation to purchase the farm-products for their own use. Initiatives of this kind — known as Community Supported Agricultural Projects (CSA's) — flourish especially in the USA. It is also possible for a group of concerned people to participate financially in the farm, for example, providing security for any new buildings that have to be built. This can lead to the creation of a close knit circle of friends surrounding a farm and allows these 'friends' to par-ticipate actively in its activities, thus acquiring a closer under-standing of farm life at all levels.

Children helping at Sturts Farm

These types of shared responsibility for the land are still at a pioneering stage, with possibilities still being sought for new forms of co-operation and social interrelationships. For example, in Britain, some farms have formed companies to sell shares and offer shareholders special discounts on farm produce, as well as the opportunity of personal involvement in the farm.[19]

$\sim$ **6** $\hookleftarrow$

Looking to the future

What are the real challenges and questions which arise for the further development of biodynamic farming?

Although this method of farming has clearly proved its worth and the feasibility and practicality of its anthroposophical starting points, it has by no means been generally accepted yet. There are still many obstacles in our society which delay the expansion of biodynamic farming and the high quality foods produced. A great deal of work still has to be carried out on many different aspects in biodynamic associations to overcome these obstacles and better publicize the methods and their results.

For the sake of clarity, it is possible to divide the areas which require particular attention (and are receiving it) into a number of categories:

    1. The scientific area, subdivided into:
        a) fundamental research (into quality, breeding and seed production);
        b) practical research.
    2. The economic area.
    3. The area of regulations and government intervention.

Research into quality

Fundamental research is aimed at the subject of an understanding of life itself. What do we mean when we talk about nourishing food? Obviously we think of food which 'contains

Examining a crop of rye

everything,' minerals as well as vitamins, fibre and all the nutritional elements such as starches, protein, fats and so on.

However, there is more to it than this. There are also life forces which are present in every living organism in a strong or weak form. The human organism will assimilate more or fewer of these life forces, depending on the food consumed. Life forces are not only important for our physical health but are also important for our moods and the way in which we inter-relate with our fellow human beings. Whether our food contains the correct life forces in sufficient strength, depends to a great extent on the way in which they were produced, and particularly on the soil. In addition, much depends on the way in which food is stored, prepared and preserved.

Although people who regularly eat biodynamic products are virtually always convinced that they taste and smell better, that

Table 4. The keeping quality of carrots, as influenced by preparations 500 and 501 (Samaras, 1977)

Keeping qualities	Control	Treatment	
		3 x 500	3 x 500 4 x 501
CO_2 in 94h (mg/kg fresh matter)	2,831	2,755	2,565
Catalase activity (μmol H_2O_2/[min. g dry matter])	397	375	346
Peroxidase (μmol GJ/[min. g dry matter])	15,992	11,051	7,591
Saccharase activity (m/U/g fresh matter)	90	98	112
Bacteria (10^6/g fresh matter)	1.07	0.45	0.43
Fungi (10^3/g fresh matter)	2.7	2.7	2.6
Loss of dry matter of mashed carrots (%)	56.1	46.6	29.2
Spoilage during 164 days (%)	28.2	23.0	20.4

the vegetables are firmer and taste better, and that people with a weak digestion can more easily digest these foods, conventional science has nevertheless lacked the analytical evidence for these claims. However, recent quantitative research into the physical composition of food has demonstrated that plants grown with the use of biodynamic preparations show an improved keeping quality as well as changes in some characteristics. Table 4, above, shows an improvement in the keeping quality of biodynamically grown carrots. Sprays applied during growth also reduce losses during storage. They reduce the production of carbon dioxide, some enzyme activity and the number of epiphytic bacteria.

Several research projects carried out in Germany and Sweden have shown that the use of spray preparations 500 and 501 increases yields in, for example, sugar beet, by 8–14 per cent, stimulating the growth of leaves by 8–26 per cent (Spiess 1979, Abele 1973.) Significant increases in yields of grains, root crops and vegetables have been demonstrated.[20] Table 5 shows increases in yields of wheat grown over four years in field trials with preparations 500 and 501.

An experiment run by the Swedish Biodynamic Institute together with the University of Uppsala, showed that the higher yield of potatoes grown on a conventional plot is reduced by losses during storage and grading. The biodynamic crop showed a higher relative content of true protein, high vitamin C content, less darkening, better preservation of taste and fewer crystallization defects, which are all indicative of better quality (see Table 6 on p.108).

If we wish to learn more about the quality of foods for living organisms such as humans or animals, in other words about the vitality and life force of different foods, we will have to develop broader based research methods. Methods have already been developed and to some extent are already used. For example, Ehrenfried Pfeiffer (1899–1961) developed the sensitive crystallization method to demonstrate the qualitative aspects of life. In this method a little of the juice or dissolved substance of the object to be examined is crystallized in a solution of copper chloride. This process is carried out under particular temperature and humidity conditions. The shapes which develop in the patterns of crystallization provide information about the object's vital properties.[21]

This method has been used for many years in anthroposophical medicine to examine patients' blood. Trained blood crystal-

Table 5. Experimental treatment of wheat with preparations 500 and 501 (Spiess, 1979)

| Year | Relative yields 1973–76 (All plots received manure) | | | |
	Control	3 x 500	3 x 500 3 x 501	3 x 500 3 x 501 *1976 4x
1973	100 (=3.0 t/ha)	109	117+	121+
1974	100 (=4.15 t/ha)	106	109	111+
1975	100 (=4.1 t/ha)	105	102	102
1976	100 (=3.0 t/ha)	105+	104	109++

Wheat

Results	*Management*	
	Conven- tional	Bio- dynamic
Yields, October (t/ha)	38.2	34.2
Losses by grading and storage	30.2	12.5
Weight in April (t/ha)	26.6	30.0
Crude protein (% of dry matter)	10.4	7.7
True protein (% of crude protein)	61.0	65.8
EAA (Oser)	58.9	62.8
Vitamin C (mg/100g fresh matter)	15.5	18.1
Darkening of extract E.10^3, 48h, 8°C	462	354
Decomposition of extract (elec. conductivity)	30.9	22.0
Crystallization defects	5.2	4.2
Taste points (best = 4) December	3.0	3.1
April	2.3	2.7
Cooking defects December	4.1	1.8
April	9.2	2.1

Table 6. Yields and quality of potatoes under conventional and biodynamic management (Pettersson, 1977).

lization researchers are able to discover a great deal of information about a patient's state of health from the appearance of these blood crystallizations. The method has also been used with regard to the quality of different foods, but the results have not yet produced any convincing or reliable information. Up to now there are insufficient facilities or finance to carry out more pioneering work in this field.

Another method which reveals the qualitative rather than the quantitative aspects is the capillary dynamolysis method (using

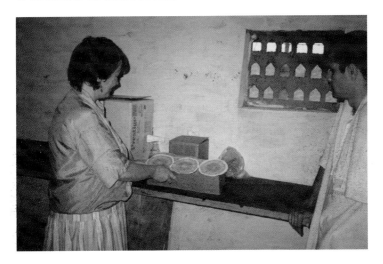

Chromatography

chromatography) developed by Lili Kolisko, following instructions of Rudolf Steiner.[22] In this method, the juice of the object to be examined is allowed to soak up on some filter paper together with a salt. The shapes which appear when this is done give an indication of the life forces active in the plant when it was harvested. This method, which still needs to be further developed, is used to investigate the quality and nature of composted manure, or the life quality of soil, in addition to its nitrogen, phosphorus and potassium content.[23]

The phenomenological research method is yet another way of acquiring an insight into the living quality of an object. For a scientist who uses the phenomenological method, the quantitative analysis is part of the evaluation of a crop. After all, the phenomenological features of the crop in all its forms and colours are not simply arbitrary. It has undergone a certain development, and if this development is carefully observed, it reveals a great deal about the interaction between the growth factors of the soil, the water and the sun, and their influence on the physical development as well as the quality of the end product.

   In other words, the phenomenon of a particular crop is part of
a whole process of other phenomena, and the end result is an
expression of this process, which also includes the influences of
the stars and planets. Therefore, experiencing this process as a
whole is the only way of gaining a true insight into the actual
phenomenon of the specific crop.
   This short description of three visual methods gives some
indication of the work and procedures in this field of fundamental research. Obviously these have not exhausted the possibilities
of research into ways of making life forces visible or quantifiable.
The whole field is still in its infancy, but it is extremely important
for our insight into the way in which the health of living organisms is affected, and into their interrelationship with their total
environment, including cosmic influences.[24]

Research into breeding and seed production

New scientific knowledge increasingly provides opportunities for the genetic manipulation of plants and animals. Biodynamic and organic farms, too, need to develop their own varieties of seeds for sowing, as well as ensure healthy lines of livestock.

Virtually everywhere in the world cows are inseminated with the sperm of a limited number of top breeding bulls. There is therefore little or no connection between the type of animal and the specific conditions of a particular farm, region, or climate.

The same applies to the seed developed for sowing. The seed is rarely produced on a farm, and this is not surprising because, except in the case of cereals where the ripe grain is harvested, other seed crops require special facilities and expertise.

The production and improvement of seed are entirely in the hands of seed producers, who sell their seed throughout the world and have a programme of seed improvement entirely geared to this. They are interested only in those varieties which

Seed production

Trial plots at Murtle in Aberdeenshire

can be used in as many regions as possible. Most seed production is owned by major petrochemical companies. Therefore the selection of new varieties is always aimed at current methods of cultivation and the use of artificial fertilizers. A great deal of use is made of so-called 'hybrid' varieties which produce highly uniform crops and a high yield. These varieties are developed every year by means of cross-fertilization and cannot be propagated further. In this way the farmers become entirely dependent on the seed producer.

Against this background, there is a real need for the development of varieties suitable for organic or biodynamic methods of cultivation. Several scientists, farmers and growers have been working in this area in recent years and, in many countries, farmers and gardeners have started to produce seed and exchange it

with one another on a small scale. A small number of biodynamic seed producers already operate commercially in Switzerland, Germany and Holland, while plans are under way for
similar enterprises in the USA and Britain. Those engaged at the
research level maintain regular contact internationally, but there
is still a great deal of practical exploration to be carried out, in
particular to gain a clearer understanding based on anthroposophical ideas, of what is actually involved when plants or animals are altered through genetic manipulation.

Another area where a great deal of work is being carried out
is related to the optimum planting of trees and shrubs. It appears
that the stellar constellations have an influence on the germination and growth of different varieties.[25] This kind of research is
carried out in nurseries and brings us to the broader area of
practical research.

Practical research

Obviously practical questions need to be answered through some
form of practical research. Examples of current research which
spring to mind relate to the use of manure, the use of the correct
mixture of grass and clover in meadows, the question of weed
control and related equipment. A number of different biodynamic institutions are working in these areas, including a number of testing stations, for example, in Germany and America.[26]
Because of the rising public interest in environmentally friendly
methods, this sort of research is also pursued in a number of
government agriculture research institutes. In recent years there
have been several government researchers working closely with
biodynamic farms.

Economic considerations

Agriculture is an activity which takes place in the economic
sector. At least this is how it is usually viewed, but is this really

a correct view? During the last few decades in particular it has become clear that there are all sorts of unforeseen side effects and negative consequences when agriculture is solely considered in compliance with general economic laws.

For virtually every industrial production activity, it goes without saying that what is produced is what sells best, what is most in demand and what leads to the highest profits. In agriculture — which, in spite of much current practice, is not an industry — this approach simply does not work. Growing potatoes year after year because this gives the best profit per hectare ends up as a disaster. The plants become diseased and so does the soil. The farmer most certainly has to take into account laws other than the laws of economics: specifically, the laws of cyclical processes, diversity and ecology. If they fail to do so, they will not survive economically in the long term; either they or their descendants will eventually reap the consequences.

This need to take into account the laws of the farm as an organism means that in sales and trading it is essential to find forms of co-operation which permit the farmer to develop the farm in a balanced way. Therefore, a situation should be developed where total dependence on a wholesaler or processor is avoided, and where there is close consultation between producers and distributors, as well as between producers themselves, in a joint attempt to establish feasible and necessary prices for the products concerned.

As discussed earlier, direct sales to the consumer form an increasingly important part of the economic picture. In order to allow organic agriculture to expand further, it is necessary for its market and sales to increase. It is essential to educate the consumer market and to publicize the quality of the products. To make this quiet voice heard against the roar of the commercial world, so that the message will reach and inform the public, demands an enormous effort.

Regulations and government intervention

Although the governments in most European countries ostensibly support a cleaner environment and better agricultural methods, and biodynamic farming is deeply concerned with these matters, it is constantly clear that all sorts of government measures actually serve as an obstacle to the practice of biodynamic farming. Nor are the members of the power bloc of the established agricultural interests prepared to compromise in any respect.

In 1983, a restriction of the annual milk lake in the EC was imposed by obliging every farmer to produce 6% less milk than in previous years. This also applied to biodynamic and organic farmers, although they had already voluntarily reduced their milk production by changing their methods. The result of the milk quotas was that farmers who did not have dairy cows, and were therefore not given a milk quota, were unable to make their farms more balanced by expanding with dairy cattle.

The legislation on manure, and all the restrictions and provisions contained in this, was not introduced for organic farmers but for intensive cattle breeders who were producing far too much manure. Nevertheless, biodynamic and organic farmers have also had to comply with the regulations.

Another problem has arisen with regard to the use of herbicides and pesticides. The government (justifiably) wishes to restrict the use of these chemicals, preferably in an EC context. For this reason, new chemicals are subject to extremely strict, time-consuming and therefore expensive control procedures, and a great deal of research information must be submitted. But this also applies, for example, to nettle tea, which has been used for centuries and is quite harmless. The result is that environmentally friendly herbicides and pesticides which are sometimes used in organic farming, are not permitted because there are no companies which can afford these expensive procedures, as the turnover of the product is so small that it will not cover the costs.

Yet another problem is that government subsidies to research which is so essential, are still extremely limited, and there is still very little understanding of its usefulness for agriculture in the future.

A growing understanding

In most countries, discussions have taken place about all these problems with government bodies, also at an international level by IFOAM, amongst others. Despite the limited nature of these discussions so far, government bodies have nevertheless revealed a growing readiness to listen to and consider the viewpoints and requirements of biodynamic and organic agriculture.

In many areas around the world, a great deal of work is being carried out to develop our understanding of different types of biodynamic and organic farming methods. These activities are aimed at a future in which the needs of the natural environment and living spiritual organisms are once again uppermost in agriculture, and not the requirements of chemical and technical methods which are limited in their terms of reference and so often applied without due heed for their consequences. For the future viability of human societies in relation to the Earth, we urgently need an agriculture which is in harmony with the natural world, as well as providing for basic human needs. We cannot do without chemistry and technology, but these should always be subordinate and subservient to the living relationship of the farming community, the landscape and nature.

References

1. For further explanation of life, or etheric forces see Chapter 1 of Steiner, *Theosophy.*
2. See Bockemühl, *Towards a Phenomenology.*
3. For further explanation of the astral body see Chapter 1 of Steiner, *Theosophy.*
4. See Steiner, *Agriculture.*
5. See Pfeiffer, *Soil Fertility, Renewal and Preservation.*
6. Virbela Flowforms are produced in various countries. They were developed by John Wilkes and his co-workers at the Virbela Flowform Research Centre (Emerson College, Forest Row, East Sussex, England) from which further information can be obtained. They have a website at www.anth.org.uk/virbelaflowforms/. See also Wilkes, *Flowforms.*
7. See Heinze and Breda, 'Versuche über Stallmistkompostierung' [Experiments on the composting of stable manure] in *Lebendige Erde*, 2, 3–10. 1962.
8. See Abele, *Untersuche des Rotteverlaufes von Guelle bei verschiedener Behandlung* [Studies on the rotting of liquid manure] Institut für biologisch-dynamische Forschung, Darmstadt 1976.
9. See Koepf, 'Experiments in treating liquid manure,' in *BioDynamics*, No 79, 1979.
10. Important research work has been published by Michael Rist at Zürich University on the proper husbandry of farm animals, enabling them to express their natural characteristics and attain high levels of health and longevity (Rist *et al., Artgemäße Nutztierhaltung*, Freies Geistesleben, Stuttgart 1987).
11. Several of these remedies are listed in Chapter 7 of Koepf, Pettersson, Schaumann, *Bio-Dynamic Agriculture.* An English translation of Spielberger, Schaette, *Biologische Stallapotheke*, [Biological Farm Animal Remedies], is in preparation.
12. See Klett, *Untersuchungen über Licht- und Schattenqualität in Relation zum Anbau und Test von Kieselpräparaten zur Qualitätshebung* [Studies of the quality of light and shadow] Institut für Biologisch-dynamische Forschung, Darmstadt 1968;
    Abele, *Vergleichende Untersuchungen zum konventionellen und biolo-*

gisch-dynamischen Pflanzenbau [Comparative studies on conventional and biodynamic plant cultivation] PhD thesis, Gießen 1973; Pettersson, 'Vergleichende Untersuchungen zum konventionellen und biologisch-dynamischen Pflanzenbau' in Lebendige Erde, 5.175–80, 1977;
Wistinghausen, Was ist Qualität? [What is Quality?] *Lebendige Erde*, Darmstadt 1979.

13. See Pfeiffer, *Biodynamic Treatment of Fruit Trees.*

14. See Steiner, *Nine lectures on Bees.*

15. For example Abele, *Vergleichende Untersuchungen* (Note 12); Maria Thun has also published many of the results of her research in the annual issue of the *Biodynamic Sowing and Planting Calendar.*

16. See Thun, *Results from the Sowing and Planting Calendar,* and the *Biodynamic Sowing and Planting Calendar.*

17. See Kolisko, *Agriculture of Tomorrow*; Fyfe, *Moon and Plant*; Steiner, *Agriculture.*

18. For example the Camphill Movement has schools, homes, villages in many different countries in Britain and Europe, as well as Southern Africa and the North America. For further information contact: Camphill Village Trust (Delrow House, Hilfield Lane, Aldenham, Watford, Herts WD2 8DJ, England).

19. See Groh, T. & McFadden S; *Farms of Tomorrow: Community Supported Farms and Farm Supported Communties.* For example Old Plawhatch Farm (Sharpthorne, West Sussex) is a developing community-owned biodynamic farm, administered by St Anthony's Trust, a registered charity. The main objectives of the farm are to provide an education for students in the principles of biodynamic agriculture, and to develop the farm according to these principles. The farm is approximately 150 acres, including 50 acres woodland, and supports 50 cows and up to 20 young stock with a few pigs. They have a farm shop selling their own vegetables, dairy produce and meat and run a green-top milk round.

Another example is Ceridwen, (Sharpham, Ashprington, Totnes, Devon) a co-operative venture involving several biodynamic enterprises, a beef and sheep farm with chickens, a market garden and an orchard.

20. See Klett, *Untersuchungen über Licht- und Schattenqualität* (Note 12 above);
Pettersson, *Vergleichende Untersuchungen* (Note 12 above);
Thun and Heinze, *Anbauversuche und Zusammenhänge zwischen Mondstellung im Tierkreis und Kulturpflanzen,* [Experiments on

connection between the position of the moon in the zodiac and the cultivation of plants] *Lebendige Erde*, Darmstadt 1973;

Spiess, 'Über die Wirkung der biologisch-dynamischen Präparate Hornmist und Hornkiesel' [On the effect of the biodynamic preparations] *Lebendige Erde* 4/5, 1979;

Abele, *Vergleichende Untersuchungen* (Note 12 above).

21. The methodical basis and progress made with this test have been reported by Krüber, *Kupferchloridkristallisation,* Weleda, Schwäbisch Gmünd 1950;

    Selawry, *Die Kupferchloridkristallisation in Naturwissenschaft und Medizin,* Fischer, Stuttgart 1957;

    Engqvist, 'Strukturveränderungen im Kupferchloridkristallisationsbild,' *Lebendige Erde*, 3, 1961;

    Engvist, 'Pflanzenwachstum in Licht und Schatten,' *Lebendige Erde,* 2, 1963;

    Engvist, *Gestaltkräfte des Lebendigen,* Klostermann, Frankfurt 1970;
    Pettersson, 'Beiträge zur Entwicklung der Kristallisationsmethode' *Lebendige Erde,* 1, 1957.

22. See Kolisko, *Agriculture of Tomorrow;* a modification of this method are the chromatograms developed by Pfeiffer, *Chromatography Applied to Quality Testing.*

23. Chromatograms have also been used to test spinach and carrots and play part in the selection of plants and harvest times for herbs used in medicines (Fyfe, *Moon and Plant).*

24. A one year course is offered in natural sciences which specialises in developing this method. This takes place at the Goetheanum, Dornach, Switzerland and is run by Dr J. Bockemühl.
    See also Bockemühl, *In Partnership with Nature.*

25. Three generations of the Schmidt family have been working on practical research into the development of stronger, more resistant strains of cereals; the zodiacal influence on tree cultivation; and the development of new livestock breeding techniques (Verein für Pflanzenzucht eV, Rittershain, 36219 Cornberg, Germany).

26. In Germany: Forschungsring für Biologisch-Dynamische Wirtschaftsweise eV, (Baumschulenweg 11, 64295 Darmstadt) co-ordinates all research done in Germany and publishes data.
    In USA: Biodynamic research carried out and on-farm research coordinated by Michael Fields Agricultural Institute (3293 Main Street, East Troy WI 53120).

Further Reading

Biodynamics, (New Zealand Bio-Dynamic Association) Random House, Auckland 1989.

Bockemühl, J, *In Partnership with Nature*, Anthroposophic Press, New York.

Bockemühl, J, (ed.), *Towards a Phenomenology of the Etheric World,* Anthroposophic Press, New York 1985.

Conford, P. *The Origins of the Organic Movement*, Floris Books, Edinburgh 2003.

Fyfe, A, *Moon and Plant*, Society for Cancer Research, Arlesheim 1967.

Groh, T, and S McFaddens, *Farms of Tomorrow: Community Supported Farms and Farm Supported Communities*, Bio-Dynamic Literature, Pennsylvania 1990.

Grotzke, H, *Biodynamic Greenhouse Management*, Biodynamic Literature, Pennsylvania 1990.

Kimberton Hills Agricultural Calendar, Kimberton Hills Publications (annual).

Koepf, H, *Compost*, Biodynamic Literature, Pennsylvania 1990.

—, The Biodynamic Farm, Anthroposophic Press, New York 1989.

Koepf, H, Bo Pettersson and W Schaumann, *Bio-Dynamic Agriculture, an Introduction.* Anthroposophic Press, New York 1976.

Kolisko, E & L, *Agriculture of Tomorrow*, Kolisko Archive Publications, Ringwood 1982.

Pfeiffer, E E, *Biodynamic Treatment of Fruit Trees, Berries and Shrubs*, Bio-Dynamic Farming and Gardening Association, Pennsylvania 1976.

—, *Chromatography Applied to Quality Testing*, Bio-Dynamic Farm and Garden Association, Pennsylvania 1960.

—, *Soil Fertility, Renewal and Preservation,* Lanthorn Press, East Grinstead 1984.

—, *Biodynamic Gardening and Farming* (3 vols), Mercury, New York 1983–84.

Podalinsky, Alex, *Bio-Dynamic Agriculture* (2 vols), Gavemer Foundation, Sydney 1985.

Steiner, Rudolf, *Agriculture,* Bio-Dynamic Agriculture Association, London 1974.

—, *Nine Lectures on Bees*, Steinerbooks, New York 1988.

—, *Theosophy*, Anthroposophic Press, New York 1994.

—, *Truth and Knowledge,* Steinerbooks, New York 1981.

Thun, Maria, *The Biodynamic Sowing and Planting Calendar*, Floris Books, Edinburgh (annual).

—, *Gardening for Life — The Biodynamic Way*, Hawthorn Press, Stroud, 1998.

—, *Results from the Biodynamic Sowing and Planting Calendar,* Floris Books, Edinburgh 2003.

Wilkes, John, *Flowforms*, Floris Books, Edinburgh 2003.

Worldwide Organizations

Australia: **Biodynamic Agricultural Association**
PO Box 54, Bellingen, NSW, 2454
Tel: 612 6655 0566
Fax: 612 6655 0565
Email: poss@midcoast.com.au
Web: www.biodynamics.net.au

Austria: **Demeter-Bund Österreich**
Hietzinger Kai 127/2/31, A-1130
Wien, Austria
Tel: 43-1879-47-01
Fax: 43-1879-47-22
Email: info@demeter.at
Web: www.demeter.at

Brazil: **IBD Instituto Biodinamico**
Rua Prudente de Morais, 530
BR-18602-060 Botucatu/Sao Paulo
Brasil
Tel/Fax: 55-1-468225066
Email: ibd@ibd.com.br
Web: www.ibd.com.br

Canada: **Demeter Canada**
115 Des Myriques, CDN-Catevale
Q. C. J0B 1W0, Canada
Tel: 1-819-843-8488
Email: laurier.chabot@sympatico.ca
Web: www.demetercanada.com

Denmark: **Demeterforbundet i**
Forening for Biod, Jordbrug
Birkum Bygade 20, DK-5220-
Odense SO, Denmark
Tel: 45-6597-3050
Fax: 45-6597-3250
Email: biodynamisk@mail.tele.dk
Web: www.biodynamisk.dk

Egypt: **Bio-Dynamic Association**
3 Belbes Desert Road, POB 1535 Alf
Maskan, ET-11777 Cairo, Egypt
Tel: 20-2656-4154
Fax: 20-2656-7828
Email: EBDA@sekem.com

Finland: **Biodynaaminen Yhdistys**
Biodynamiska Foereningen
Uudenmaankatu 25 A 4,
FIN-00120 Helsinki 12, Finland
Tel: 35-89-644160
Fax: 35-89-6802591
Email: info@biodyn.fi
Web: www.biodyn.fi

France: **Association Demeter**
5 Place de la Gare, F-68000
Colmar, France
Tel/Fax: 33 389 414 395
Email: demeter@pandemonium.fr

Germany: **Demeter-Bund e.V.**
Brandschneise 1, D-64295
Darmstadt, Germany
Tel: 49-6155-8469-0
Fax: 49-6155-8469-11
Email: Info@Demeter.de
Web: www.demeter.de

Ireland: **Biodynamic Agricultural Association**
The Watergarden,
IRL-Thomastown,
Co. Kilkenny
Tel/Fax: 35 35654214
Email: bdaai@indigo.ie
Web: www.demeter.ie

Italy: Demeter Associazione per la Tutela della Qualitá
Biodinamica in Italia, Strada, Naviglia 11/A, I-43100 Parma, Italy
Tel: 39-0521-7769-62
Fax: 39-0521-7769-73
Email: demeter.italia@tin.it
Web: www.demeter.it

Luxemburg: Veräin für biologisch dyn. Landwirtschaft
Letzeburg a.s.b.l., Demeter Bond, Letzebuerg 13, parc d'activité Sydrall L-5365 Munsbach, Luxembourg
Tel: 352-261 533-80 /Fax 81
Email: demeter@pt.lu
Web: www.demeter.lu

Netherlands: Vereniging voor Biologisch-Dynamische Landbouw en Voeding
Postbus 236, NL-3970 AE Driebergen, The Netherlands
Tel: 31-343-531740
Fax: 31-343-516943
Email: Info@demeter-bd.nl
Web: www.demeter-bd.nl

New Zealand: Biodynamic Association
PO Box 39045, Wellington Mail Centre, New Zealand.
Tel: 64 458 953 66/ Fax: /65
Email: biodynamics@clear.net.nz
Web: www.biodynamic.org.nz

Norway: Debio
N-1940 Björkelangen, Norway
Tel: 47-63-862-670
Fax: 47-63-856-985
Email: kontor@debio.no
Web: www.debio.no

South Africa: Biodynamic and Organic Agricultural Association
P.O. Box 115, ZA-2056 Paulshof Gauteng
Tel/Fax: 27-118-0371-91
Email: eleanor@pharma.co.za

Sweden: Stiftelsen Biodynamiska
Skillebyholm, S-15391 Järna, Sweden
Tel: 0046 8 551 577 02
Fax: 0046 8 551 577 81
Email: sbfi@jdb.se
Web: www.jdb.se/sbfi/english

Switzerland: Demeter Verband Schweiz
Stollenrain 10, Postfach 344 CH-4144 Arlesheim, Switzerland
Tel: 41-61-706-96-43
Fax: 41-61-706-96-44
Email: info@demeter.ch
Web: www.demeter.ch

UK: Biodynamic Agricultural Association (BDAA),
The Secretary, (BDAA), Painswick Inn, Stroud, Glos., UK.
Tel/Fax: 01453 759501
Email: bdaa@biodynamic.freeserve.co.uk
Web: www.biodynamic.org.uk

USA: Biodynamic Farming and Gardening Association
25844 Butler Road, Junction City, OR 97448
Tel: 001 888 5167797
Fax: 001 541 9980106
Email: biodynamic@aol.com
Web: www.biodynamics.com